Smart Sustainable Manufacturing Systems

Smart Sustainable Manufacturing Systems

Special Issue Editors

Dimitris Kiritsis (Kyritsis)
Gökan May

MDPI • Basel • Beijing • Wuhan • Barcelona • Belgrade

MDPI

Special Issue Editors
Dimitris Kiritsis (Kyritsis)
ICT for Sustainable Manufacturing,
EPFL SCI STI DK, Lausanne, Switzerland

Gökan May
ICT for Sustainable Manufacturing,
EPFL SCI STI DK, Lausanne, Switzerland

Editorial Office
MDPI
St. Alban-Anlage 66
4052 Basel, Switzerland

This is a reprint of articles from the Special Issue published online in the open access journal *Applied Sciences* (ISSN 2076-3417) from 2018 to 2019 (available at: https://www.mdpi.com/journal/applsci/special_issues/Sustainable_Manufacturing)

For citation purposes, cite each article independently as indicated on the article page online and as indicated below:

LastName, A.A.; LastName, B.B.; LastName, C.C. Article Title. *Journal Name* **Year**, *Article Number, Page Range.*

ISBN 978-3-03921-201-9 (Pbk)
ISBN 978-3-03921-202-6 (PDF)

Contents

About the Special Issue Editors

Dimitris Kiritsis (Kyritsis) (Prof. Dr.) is Faculty Member at the Institute of Mechanical Engineering of the School of Engineering of EPFL, Switzerland, where he leads a research group in ICT for Sustainable Manufacturing. He serves also as Director of the Doctoral Program of EPFL on Robotics, Control, and Intelligent Systems (EDRS). His research interests are Closed Loop Lifecycle Management, IoT, Semantic Technologies, and Data Analytics for Engineering Applications. He served also as Guest Professor at the IMS Center at the University of Cincinnati, and Invited Professor at the University of Technology of Compiègne, University of Technology of Belfort-Montbéliard, and ParisTech ENSAM Paris. Prof. Kiritsis is actively involved in EU research programs in the area Factories of the Future and Enabling ICT for Sustainable Manufacturing and over 220 publications. Dimitris has served as Chair of IFIP WG5.7 Advanced Production Management Systems since September 2013. From 2013 to 2017 he was member of the Advisory Group of the European Council on Leadership on Enabling Industrial Technologies (AG LEIT-NMBP). He is also a founding fellow member of the International Society for Engineering Asset Management (ISEAM) and of various international scientific communities in his area of interests, including EFFRA, and is among the initiators of the IOF (Industrial Ontologies Foundry).

Gökan May is a Postdoctoral Researcher in Mechanical Engineering Department at École Polytechnique Fédérale de Lausanne, a Fellow of the World Economic Forum's Global Future Council on Production, and Editorial Board Member of the World Manufacturing Forum. He received his Ph.D. in 2014 (Summa Cum Laude) and M.Sc. degree in 2010 (with Highest Honours) in Management, Economics and Industrial Engineering from Politecnico di Milano. During 2013, he worked as a Researcher at Pennsylvania State University in the Department of Industrial and Manufacturing Engineering. His major fields of research in the main area of data-driven smart and sustainable manufacturing include energy efficient manufacturing, product and asset lifecycle management, zero-defect manufacturing, designing human-centric workplaces of the future, and eco-efficient design of production facilities. Dr. Gökan May has published more than 30 papers in international journals and conference proceedings, and has been involved in several European Horizon 2020 and FP7 FoF and ICT proposals and projects, including QU4LITY, Z-BRE4K, BOOST4.0, Z-Fact0r, Man-Made, PLANTCockpit, EMC2 Eco-Factory, and LeanPPD.

applied
sciences

MDPI

Editorial

Special Issue on Smart Sustainable Manufacturing Systems

Gökan May * and Dimitris Kiritsis *

École Polytechnique Fédérale de Lausanne, ICT for Sustainable Manufacturing, EPFL SCI-STI-DK, Station 9, 1015 Lausanne, Switzerland
* Correspondence: gokan.may@epfl.ch (G.M.); dimitris.kiritsis@epfl.ch (D.K.)

Received: 24 May 2019; Accepted: 27 May 2019; Published: 31 May 2019

1. Introduction

With the advent of disruptive digital technologies, companies are facing unprecedented challenges and opportunities. Advanced manufacturing systems are of paramount importance in making key enabling technologies and new products more competitive, affordable and accessible as well as fostering their economic and social impact. The manufacturing industry also serves as an innovator for sustainability since automation coupled with advanced manufacturing technologies have helped to transition manufacturing practices to the circular economy [1]. In this context, shifting paradigms comprehend the 360-degree makeover of factories, from shop-floor to supply chain, from blue collar staff to top management, from employee to stakeholder. In that regard, the objective of smart and sustainable manufacturing systems of the future is to enable clean and competitive manufacturing systems irrespective of factories' location or size, and to find opportunities based on sustainability issues to grow beyond their borders [2]. To that end, this special issue of the journal Applied Sciences devoted to the broad field of Smart Sustainable Manufacturing Systems was introduced to explore recent research into the concepts, methods, tools and applications for smart sustainable manufacturing in order to advance and promote the development of modern and intelligent manufacturing systems.

2. Smart Sustainable Manufacturing Systems

In light of the above, this special issue collects the latest research on relevant topics, and addresses present challenging issues with the introduction of smart sustainable manufacturing systems. There were 24 papers submitted to this special issue, and 9 papers were accepted (i.e., a 37.5% acceptance rate). Various topics have been addressed in this special issue, mainly on design of sustainable production systems and factories; industrial big data analytics and cyber physical systems; intelligent maintenance approaches and technologies for increased operating life of production systems; zero-defect manufacturing strategies, tools and methods towards on-line production management; and connected smart factories.

The first paper, Production and Maintenance Planning for a Deteriorating System with Operation-Dependent Defectives, authored by H. Rivera-Gómez, O. Montaño-Arango, J. Corona-Armenta, J. Garnica-González, E. Hernández-Gress, and I. Barragán-Vite provides new insights into the area of sustainable manufacturing systems by analysing the novel paradigm of integrated production logistics, quality and maintenance design, and investigates the optimal production and maintenance switching strategy of an unreliable deteriorating manufacturing system. The paper presents a model that defines the joint production and maintenance switching strategies minimizing the total cost over an infinite planning horizon [3].

The second paper, Kernel-Density-Based Particle Defect Management for Semiconductor Manufacturing Facilities, proposes a particle defect management method for the reduction of the defect ratio in semiconductor manufacturing facilities, and presents a kernel-density-based particle map that can overcome the limitations of the conventional method [4], authored by S. Park, S. Kim, and J.-G. Baek.

The third paper of the Special Issue, A Multi-Usable Cloud Service Platform: A Case Study on Improved Development Pace and Efficiency, authored by J. Lindström, A. Hermanson, F. Blomstedt, and P. Kyösti, addresses a micro small and medium-sized enterprise (SME) in Sweden and its journey of developing and operating a multi-usable cloud service platform for big data collection and analytics [5].

The article An Integrated Open Approach to Capturing Systematic Knowledge for Manufacturing Process Innovation Based on Collective Intelligence by G. Wang, Y. Hu, X. Tian, J. Geng, G. Hu, and M. Zhang [6] builds a novel holistic paradigm of process innovation knowledge capture based on collective intelligence as a foundation for the future knowledge-inspired computer-aided process innovation and smart process planning.

The next two articles focus on approaches and methods on supply chain level. The first one, A Modified Method for Evaluating Sustainable Transport Solutions Based on AHP and Dempster–Shafer Evidence Theory by L. Chen and X. Deng, presents a transport sustainability index (TSI) as a primary measure to determine whether transport solutions have a positive impact on city sustainability [7]. The subsequent paper, Dynamic Supply Chain Design and Operations Plan for Connected Smart Factories with Additive Manufacturing authored by B. Chung, S.I. Kim and J.S. Lee suggests a general planning framework and various optimization models for dynamic supply chain design and operations plan [8].

The seventh paper in this Special Issue, The Role of Managerial Commitment and TPM Implementation Strategies in Productivity Benefits written by J. Díaz-Reza, J. García-Alcaraz, L. Avelar-Sosa, J. Mendoza-Fong, J. Sáenz Diez-Muro, and J. Blanco-Fernández [9], proposes a structural equation model to integrate four latent variables: Managerial commitment, preventive maintenance, total productive maintenance, and productivity benefits.

Subsequently, Opportunities for Industry 4.0 to Support Remanufacturing by S. Yang, A.M.R., J. Kaminski, and H. Pepin reviews the challenges encountered by the remanufacturing sector and discusses how the Industry 4.0 revolution could help to effectively address these issues and unlock the potential of remanufacturing [10].

Last but not least, the final article Hybrid Laminate for Haptic Input Device with Integrated Signal Processing of R. Schmidt, A. Graf, R. Decker, V. Kräusel, W. Hardt, D. Landgrebe, and L. Kroll presents a new tool concept for joining and forming hybrid laminates in a manufacturing process [11].

3. Future Research

Although the special issue has been closed, more in-depth research in smart sustainable manufacturing systems is expected. In particular, demonstrative scenarios that pertain to smart design, smart machining, smart control, smart monitoring, and smart scheduling to highlight key enabling technologies and their possible applications to Industry 4.0 smart manufacturing systems could complement the research aspects covered within this Special Issue.

Acknowledgments: We would like to take this opportunity to thank all the authors, reviewers, and dedicated editorial team of *Applied Sciences*. The special issue would not have been possible without the contributions and generous support of them.

Conflicts of Interest: The authors declare no conflict of interest.

References

1. 2018 World Manufacturing Forum Report, Recommendations for the Future of Manufacturing. Available online: https://www.worldmanufacturingforum.org/report (accessed on 24 May 2019).
2. May, G.; Stahl, B.; Taisch, M. Energy management in manufacturing: Toward eco-factories of the future—A focus group study. *Appl. Energy* **2016**, *164*, 628–638. [CrossRef]

3.   Rivera-Gómez, H.; Montaño-Arango, O.; Corona-Armenta, J.; Garnica-González, J.; Hernández-Gress, E.; Barragán-Vite, I. Production and Maintenance Planning for a Deteriorating System with Operation-Dependent Defectives. *Appl. Sci.* **2018**, *8*, 165. [CrossRef]

4.   Park, S.; Kim, S.; Baek, J.-G. Kernel-Density-Based Particle Defect Management for Semiconductor Manufacturing Facilities. *Appl. Sci.* **2018**, *8*, 224. [CrossRef]

5.   Lindström, J.; Hermanson, A.; Blomstedt, F.; Kyösti, P. A Multi-Usable Cloud Service Platform: A Case Study on Improved Development Pace and Efficiency. *Appl. Sci.* **2018**, *8*, 316. [CrossRef]

6.   Wang, G.; Hu, Y.; Tian, X.; Geng, J.; Hu, G.; Zhang, M. An Integrated Open Approach to Capturing Systematic Knowledge for Manufacturing Process Innovation Based on Collective Intelligence. *Appl. Sci.* **2018**, *8*, 340. [CrossRef]

7.   Chen, L.; Deng, X. A Modified Method for Evaluating Sustainable Transport Solutions Based on AHP and Dempster–Shafer Evidence Theory. *Appl. Sci.* **2018**, *8*, 563. [CrossRef]

8.   Chung, B.; Kim, S.I.; Lee, J.S. Dynamic Supply Chain Design and Operations Plan for Connected Smart Factories with Additive Manufacturing. *Appl. Sci.* **2018**, *8*, 583. [CrossRef]

9.   Díaz-Reza, J.; García-Alcaraz, J.; Avelar-Sosa, L.; Mendoza-Fong, J.; Sáenz Diez-Muro, J.; Blanco-Fernández, J. The Role of Managerial Commitment and TPM Implementation Strategies in Productivity Benefits. *Appl. Sci.* **2018**, *8*, 1153. [CrossRef]

10.   Yang, S.; MR, A.; Kaminski, J.; Pepin, H. Opportunities for Industry 4.0 to Support Remanufacturing. *Appl. Sci.* **2018**, *8*, 1177. [CrossRef]

11.   Schmidt, R.; Graf, A.; Decker, R.; Kräusel, V.; Hardt, W.; Landgrebe, D.; Kroll, L. Hybrid Laminate for Haptic Input Device with Integrated Signal Processing. *Appl. Sci.* **2018**, *8*, 1261. [CrossRef]

*applied
sciences*

MDPI

Article

Production and Maintenance Planning for a Deteriorating System with Operation-Dependent Defectives

Héctor Rivera-Gómez *, Oscar Montaño-Arango, José Ramón Corona-Armenta,
Jaime Garnica-González, Eva Selene Hernández-Gress and Irving Barragán-Vite

Academic Area of Engineering, Autonomous University of Hidalgo, Pachuca-Tulancingo Road km. 4.5,
City of Knowledge, Mineral de la Reforma 42184, Hidalgo, Mexico; omontano@uaeh.edu.mx (O.M.-A.);
jrcorona@uaeh.edu.mx (J.R.C.-A.); jgarnica@uaeh.edu.mx (J.G.-G.); evah@uaeh.edu.mx (E.S.H.-G.);
irvingb@uaeh.edu.mx (I.B.-V.)
* Correspondence: hriver06@hotmail.com; Tel.: +52-771-712-000 (ext. 4001)

Received: 12 December 2017; Accepted: 19 January 2018; Published: 24 January 2018

Abstract: This paper provides new insights to the area of sustainable manufacturing systems at analyzing the novel paradigm of integrated production logistics, quality, and maintenance design. For this purpose, we investigate the optimal production and repair/major maintenance switching strategy of an unreliable deteriorating manufacturing system. The effects of the deterioration process are mainly observed on the failure intensity and on the quality of the parts produced, where the rate of defectives depends on the production rate. When unplanned failures occur, either a minimal repair or a major maintenance could be conducted. The integration of availability and quality deterioration led us to propose a new stochastic dynamic programming model where optimality conditions are derived through the Hamilton-Jacobi-Bellman equations. The model defined the joint production and repair/major maintenance switching strategies minimizing the total cost over an infinite planning horizon. In the results, the influence of the deterioration process were evident in both the production and maintenances control parameters. A numerical example and an extensive sensitivity analysis were conducted to illustrate the usefulness of the results. Finally, the proposed control policy was compared with alternative strategies based on common assumptions of the literature in order to illustrate its efficiency.

Keywords: operations management; production planning; quality; deteriorating systems; maintenance

---

1. Introduction

Manufacturing systems rarely perform exactly as expected and predicted, since they may experience disorders from many unexpected events, such as equipment break-downs, delays, defectives, deterioration, etc., as reported in Liberopoulos et al. [1]. In this context, quality, production planning and maintenance define the fundamental functions to achieve success in the manufacturing industry, implying resource efficiency along the product, process, and production system life-cycle. Therefore, integrated operations management approaches are needed to have a global vision of the company by taking into account the interactions between the different key functions. This paper aims to develop an effective method for decision making on industrial strategies under integrated approach of the production control, quality, and maintenance planning.

There is a broad variety of practical problems dealing with the association between production and quality. It is clear that equipment availability, product quality, and productivity are strongly interrelated. However, these fields have been traditionally treated by manufacturers and researchers almost in isolation. Some authors have proposed frameworks for the joint production-quality relationship as in Colledani and Tolio [2], who presented an analytical method for evaluating the performance of production systems, jointly considering quality and production performance indices. Yedes et al. [3] studied a production unit that randomly shifts from an in-control to an out-of control state, where at the end of the production cycle, maintenance activities are performed depending on the state of the unit. The simulation work proposed by Rivera-Gómez et al. [4] addressed the problem of an unreliable manufacturing system that produces conforming and non-conforming parts, where due to the wear of the system, the authors considered the use of external production to supplement the limited production capacity. An algorithm integrating production and quality issues were presented by Mhada et al. [5], where they determined the buffer sizing and inspection positioning problem of large production lines, identifying promising locations for the inspection stations. Recently, Bouslah et al. [6] investigated the joint design and optimization of a continuous sampling plan, make-to-stock production and maintenance. They defined the number of successive items clear of defects required to discontinue rigorous inspection, the fraction of product sampling, the maintenance period and the amount of inventory needed as protections against disruptions. According to these studies, the relation between production and quality exists in several ways. Nevertheless, the model developed in our paper is different, because we take into account the fact that production at high rates accelerates the machine degradation and thus increases the total cost of repairs, defectives, production, etc. Therefore, the decisions involved in our formulation seek how to balance production, quality, and maintenances activities for efficient operations management.

The issues related to the maintenance of manufacturing systems are relevant to our research because in modern production systems their components are usually unreliable and so maintenance decisions should be integrated in the decision-making to properly estimate their global effect, as in the work of Mifdal et al. [7]. Who developed a method to find the optimal production rate for a manufacturing system, which produces several products in order to satisfy random demands; also, they established an economical scheduling for preventive maintenance. The study of Khatab et al. [8] addressed the problem of a production system that is continuously monitored and subject to stochastic degradation. To assess such degradation, the system undergoes preventive maintenance whenever its reliability reaches an appropriate value. Hajej et al. [9] study a manufacturing system composed by a failure prone-machine, a manufacturing store, and a purchase warehouse with service level, where a preventive maintenance plan is provided in order to decrease the failure rate. In the study of Askri et al. [10], the authors dealt with a preventive maintenance strategy and the determination of an economical production plan. Their model defines the optimal maintenance interval at which machines are maintained simultaneously. A common feature of the above papers is that they have mainly studied the joint production scheduling and maintenance planning, which has received much attention in the literature, but this does not necessarily lead to an optimal solution. Since they have disregarded the importance of quality aspects in their results. Hence, taking into account the interrelations between production, quality, and maintenance, traditionally approaches may be modified.

In the context of deteriorating systems, machine failure is probably one of the most frequently observed disruption that does deteriorate the system performance. A considerable amount of research has been spurred to address time-dependent failures. However, in most manufacturing systems is often more realistic to assume that machine reliability does depend on the degree of utilization of the machine. Thus, operation-dependent failures are common in such systems, and this assumption renders the problem much more involved, as indicated by Martinelli et al. [11], who provided the structure of a policy minimizing the long-term average backlog and inventory cost for an unreliable machine, where the failure rate is a piecewise constant function of the production rate. In the same vein, Dahane et al. [12] dealt with the problem of dependence between production and failure rates in the context of single randomly failing and repairable manufacturing system producing two products. Haoues et al. [13] were interested in the study of a production unit that aims to satisfy the deterministic market demands for multiple products. They considered that the production cost depends on the using rate of the machine, and that such machine deteriorates with increased use. Other researchers have treated the problem of production-dependent failure rates, for example Kouedeau et al. [14] analyzed a manufacturing system comprising parallel machines with failure rate depending on their productivity. They determined the productivity of the main and the supporting machine. From the discussed papers, it is evident that models considering operation-dependent failures, are rarely studied in the literature, and their focus have been mainly on the dependence between production and failure rate. Thus, one drawback of these papers is that the connection between productivity and quality deterioration have not been considered. In contrast, since deterioration is a common industrial phenomenon, our model aims to extend the concept of deterioration to state that indeed production at higher rates accelerates the machine degradation, and their effect not only may increase the failure rate but also may decline product quality.

Our research aims to generalize previous assumptions and extent several conjectures reported in the literature. In particular, we extend the work of Martinelli et al. [11], Hajej et al. [9], and Kouedeau et al. [14] in several directions: (i) at presenting an integrated production-maintenance-quality approach which serves to analyze the interactions between these three key functions; (ii) at studying the impact of a double deterioration process with continuous deterioration of part quality and reliability; (iii) at considering the dependence between productivity and product quality, leading to define operation-dependent defectives. We note that these set of characteristics have not been treated simultaneously in the literature yet. We developed a stochastic optimal control model to determine the structure of the control policy. Moreover, the obtained results are examined thorough an extensive sensitivity analysis.

The remainder of the paper is organized as follows: in the next section, the industrial motivation of the paper is presented. Section 3 describes the notation and formulation of the proposed model. In Section 4, the optimization method that is applied is detailed. The obtained joint control policy is presented in Section 5. A sensitivity analysis is carried out in Section 6. A comparative study is conducted in Section 7. Some managerial implications are discussed in Section 8. Finally, conclusions and future scope of research are provided in Section 9.

2. Industrial Motivation

In a manufacturing environment, it exits a vast number of potential disruptions that negatively affect the system's performance such as failures, wear, shortages, defectives, etc. Among these disruptions, machine failures are the most frequent problem observed in manufacturing systems. Furthermore, more realistic models are conceived at considering that machine reliability does depend on the degree of utilization of the machine leading to define operation-dependent failures, as indicated by Dong-Ping [15]. Although, such type of failures is common in production systems, they are rarely considered by researchers and practitioners. Additionally, during the last years the focus has been on the dependence between productivity and the failure rate, and just some works have studied the connection between operation-dependent failures and deterioration, as in Kouedeau et al. [14]. Nevertheless, in modern production systems, deterioration is a common industrial phenomenon. Hence, this observation raises the question of whether at considering a deterioration process, the production at higher rates may accelerate the machine degradation, indicating a dependence between deterioration and several system's performance indices such as product quality, reliability, safety, etc. Conversely, producing at low rates may contribute to an increase of shortages and incur economic losses. Thus, given the dependency of the involved cost and productivity, a trade-off is implied, and it would be advantageous to reduce the production rate from its maximum value to a more profitable level to reduce for instances the increase of defective units and failures.

The model presented in this paper has many applications especially in industries characterized by deterioration, where the production system is subject to random failures and repairs, defective quality is present and their production rates can be controlled. In particular, in situations where the production system deteriorates over time such as the automotive sector, pharmaceutical, semiconductor industries, etc.

3. Notation and Problem Statement

In this section, we define the notation used in the model formulation, also we define the manufacturing system under analysis.

3.1. Notations

The proposed model is based on the following notations:

$x(t)$	Inventory level at time t
$a(t)$	Age of the machine at time t
$u(t)$	Production rate at time t
$\xi(t)$	Stochastic process
u_{max}	Maximal production rate
u_i	Productivities of the machine
$w(t)$	Control variable for the repair/major maintenance policy at time t
$\beta(\cdot)$	Rate of defectives
d	Constant demand rate of products
Ω	Set of states of the machine
ρ	Discount rate
π_i	Limiting probability at mode i
$\lambda_{\alpha\alpha'}(\cdot)$	Transition rate from mode α to mode α'
$g(\cdot)$	Instantaneous cost function
$J(\cdot)$	Expected discounted cost function
$v(\cdot)$	Value function
τ	Jump time of $\xi(t)$
c^+	Inventory holding cost/units/time units
c^-	Backlog cost/units/time units
c_r	Repair cost
c_m	Major maintenance cost
c_d	Cost of defectives
c_{pro}	Cost of production per unit of produced parts
θ	Adjustment parameter for the rate of defectives

3.2. Problem Description

The manufacturing system under study consists of a single machine producing one part type. Nonetheless, the machine is unreliable and is subject to random events such as failures and maintenances actions of random duration. The machine can produce at diverse capacities to satisfy a constant product demand. Additionally, our considered systems has two principal features, where its failure rate increases in function of its level of deterioration, and the quality of the items produced is not perfect, there is a rate of non-conforming units. Such rate of defectives depends on the productivity of the system, thus defining a productivity-dependent defectives rate. Therefore, the system deteriorates with age and its production pace. These assumptions are common in production management. The stock is a mixture of flawless and defective product and serves as protections against shortages. To cope with the effects of the deterioration process, when the machine is at failure a fundamental problem of the decision-maker is to decide between the conduction of:

i. A minimal and inexpensive repair that serves to operate the machine for a while, but with the disadvantage that it does not restore the effect of deterioration, it leaves the machine in as-bad-as-old conditions, ABAO.

ii. An expensive major maintenance, which mitigates completely the effects of deterioration, leaving the machine in as-good-as-new-conditions, AGAN.

We intend to determine an optimal control policy that defines the appropriate production pace and the repair/major maintenance switching strategy that minimizes the average total cost comprising the inventory, backlog, defectives, production, and maintenance cost. Figure 1, illustrates the block diagram of the manufacturing system under analysis.

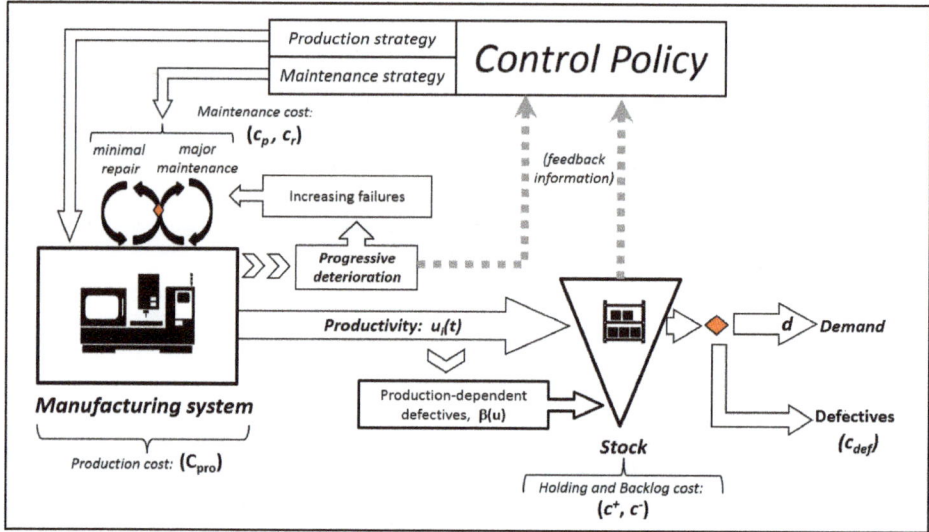

Figure 1. Manufacturing system under study.

3.3. Problem Formulation

We start by conjecturing that the manufacturing system analyzed in this paper consists of an unreliable machine subject to a double deterioration process producing a single part type. The machine mode is described by the stochastic process $\xi(t) \in \Omega = \{1, 2, 3, 4\}$. More precisely, the machine is available when it is operational ($\xi(t) = 1$), an unavailable when it is at failure ($\xi(t) = 2$). Once at failure, the decision-maker must decide between two types of maintenance actions available. When ($\xi(t) = 3$), a minimal repair is conducted where the machine has the same failure rate as before failure, in other words, it restores the system to ABAO conditions. Furthermore, when ($\xi(t) = 4$) a major maintenance is performed mitigating all the effects of the deterioration process, thus restoring the system to AGAN conditions. The transitions rates of the system $\lambda_{\alpha\alpha'}$, from state α to α', are statistically described by the state probabilities:

$$P[\xi(t + \delta t) = \alpha \mid \xi(t) = \alpha', x(t) = x, a(t) = a]$$
$$= \begin{cases} \lambda_{\alpha\alpha'}(\cdot)\delta t + o(x, a, \delta t) & \text{if } \alpha \neq \alpha' \\ 1 + \lambda_{\alpha\alpha'}(\cdot)\delta t + o(x, a, \delta t) & \text{if } \alpha = \alpha' \end{cases} \tag{1}$$

with

$$\lim_{\delta t \to 0} \frac{o(\delta t)}{\delta t} = 0; \quad \lambda_{\alpha\alpha'}(\cdot) = -\sum_{\alpha \neq \alpha'} \lambda_{\alpha\alpha'}(\cdot) \tag{2}$$

$$\lambda_{\alpha\alpha'}(\cdot) \geq 0, \quad (\alpha \neq \alpha'), \quad \forall \alpha, \alpha' \in \Omega. \tag{3}$$

The stochastic process defines a generator matrix $Q(\cdot) = (\lambda_{\alpha\alpha'}(\cdot))$, which is defined as follows:

$$Q(\cdot) = \begin{bmatrix} \lambda_{11} & \lambda_{12}(a) & 0 & 0 \\ 0 & \lambda_{22} & 0 & \lambda_{34} \\ \lambda_{31} & 0 & \lambda_{33} & 0 \\ \lambda_{41} & 0 & 0 & \lambda_{44} \end{bmatrix}. \tag{4}$$

The transition diagram of the system is presented in Figure 2.

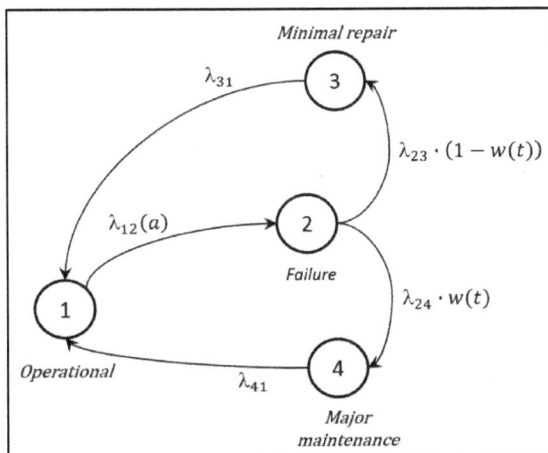

Figure 2. Transition diagram.

The transition rate $\lambda_{12}(a)$ implies that the failure rate of the machine depends on its age. The rate λ_{23} defines the transition from the failure mode to the minimal repair mode. The inverse $[\lambda_{23}\cdot(1 - w(t))]$ represents the expected delay between a call for the technician and his arrival. A similar delay is represented by the reciprocal $[\lambda_{24}\cdot w(t)]$ when the machine is send to major maintenance. Transitions λ_{31} and λ_{41} implies that the maintenance durations are defined by an exponential random variable with constant mean. Additionally, we define a binary variable $w(t) \in \{0,1\}$ that allows us to properly synchronize the transitions to the maintenance options available, as denoted in the following expression:

$$\omega(t) = \begin{cases} 0 & if\ minimal\ repair\ is\ performed \\ 1 & if\ major\ maintenance\ is\ conducted \end{cases}. \tag{5}$$

One noteworthy feature of the model is the assumption of production-dependent defectives, which implies that when the machine operates at a higher production rate, it is more likely to deteriorate faster, generating more defectives. Hence, to make this more precise, we state that the defectives rate $\beta(\cdot)$ depends on the production rate $u(t)$ according to the following expression:

$$\beta(u(t)) = \begin{cases} b_1 & if\ u(t) < u_1 \\ b_2 & if\ u(t) \in (u_1, u_2] \\ \cdots & \cdots \\ b_k & if\ u(t) \in (u_{k-1}, u_k] \\ \cdots & \cdots \\ b_n & if\ u(t) \in (u_{n-1}, u_{max}] \end{cases} \tag{6}$$

with $b_n \geq \ldots \geq b_2 \geq b_1$, and $0 \leq u_1 \leq u_2 \leq \ldots \leq u_{max}$. Where b_k and u_k are given constants. More precisely, the value of constants b_k of the defectives rate has the general form, (Kouedeu et al. [14]):

$$b_k = \eta_0 \left(\frac{u_k}{u_{max}} \right)^{\eta_1} \tag{7}$$

where η_0 and η_1 are known positive constants and u_{max} is the maximum production rate. Equation (7) serves to define the value of constant b_k. Figure 3 presents the trend of the rate of defectives $\beta(\cdot)$ for different values of η_0 and η_1. We can observe in Figure 3, the considerable influence of the productivity of the machine on the rate of defectives.

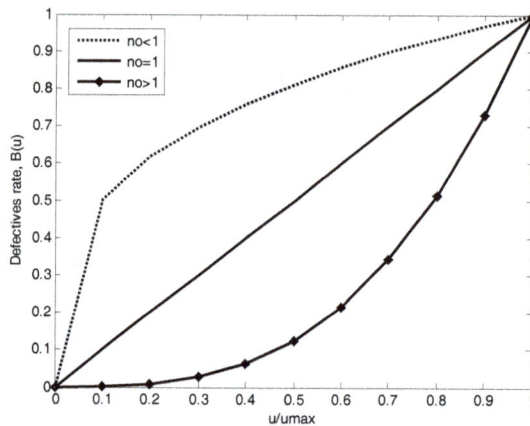

Figure 3. Defectives rate.

At considering the presence of defectives, the evolution of the stock level $x(t)$ is defined by the following differential equation:

$$\frac{dx(t)}{dt} = u(t) - d[1 + \beta(u(t))], \quad x(0) = x_0. \tag{8}$$

The constant x_0 defines the initial stock level and d denotes the demand rate. Concerning the evolution of the age of the machine $a(t)$, it implies an increasing function of the number of parts produced, and it is defined by the next differential equation:

$$\frac{da(t)}{dt} = ku(t) \tag{9}$$

$$a(T) = 0 \tag{10}$$

with k as a given positive constant and T is the last restart time of the machine. Furthermore, bearing in mind that the deterioration process also has an effect on the reliability of the machine, in particular in its failure rate $\lambda_{12}(\cdot)$. Then the lifetime distribution of a new machine follows an increasing function, as in Love et al. [16]:

$$\lambda_{12}(a(t)) = \lambda_1 + \lambda_2[1 - e^{-r\theta \, a(t)^{\eta_2}}] \tag{11}$$

where the parameter θ is useful to adjust the trend of the failure rate and $0 \leq \theta \leq 1$, λ_1 is the failure rate in AGAN conditions, λ_2 is the limit considered in the deterioration process for the rate $\lambda_{12}(\cdot)$, r, and η_2 are non-negative constants. At selecting appropriate values for r, Equation (11) can model increasing functions similar to the Weibull distribution. We present in Figure 4, the trend of the failure rate for different values of the adjustment parameter θ .

The decision variables of the model are the production rate $u(t)$, and the maintenance switching strategy $\omega(t)$. Thus, the set of feasible control policies $\Gamma(\alpha)$, including $(u(t), \omega(t))$ is given by:

$$\Gamma(\alpha) = \left\{ (u(t), \omega(t)) \in R^2, \quad 0 \leq u(t) \leq u_{max}, \quad \omega(t) \in \{0, 1\} \right\}. \tag{12}$$

We are now able to define the cost rate of the model as:

$$g(\alpha, x, a) = c^+ x^+ + c^- x^- + c_d \cdot \beta(u(t)) \cdot d + c_{pro} \cdot u(t) + c_r \cdot Ind\{\xi(t) = 3\} + c_m \cdot Ind\{\xi(t) = 4\} \tag{13}$$

with:

$$x^+ = max(0, x)$$

$$x^- = max(-x, 0)$$

$$Ind(\xi(t) = \alpha) = \begin{cases} 1 & \text{if } \xi(t) = \alpha \\ 0 & \text{otherwise} \end{cases}$$

where the cost parameters c^+ and c^- are used to penalize inventory and backlog, respectively. The parameter c_d denotes the defective cost originated by the additional handling and inspection, c_{pro} denotes the production cost, c_r is the minimal repair cost and c_m is the major maintenance cost. The objective in our model implies the determination of the optimal control policies that minimizes the integral of the following expected discounted cost:

$$v(\alpha, x, a) = \inf_{(u(t), \omega(t)) \in \Gamma(\alpha)} E\left[\int_0^\infty e^{-\rho t} g(\cdot) dt \mid \alpha(0), \ x(0), \ a(0) \right] \tag{14}$$

where ρ denotes a positive discounted rate, and $v(\cdot)$ defines the value function of the model. Based on the *optimality principle*, and at defining the cost-to-go function as $v(\cdot, t)$, we can break-up the integral of Equation (14) as follows:

$$v(\alpha, x, a, t) = \inf_{\substack{u(t), \omega(t) \\ 0 \leq t \leq \infty}} E\left[\int_0^t e^{-\rho t} g(\cdot) dt + \int_t^\infty e^{-\rho t} g(\cdot) dt \mid \alpha(0), \ x(0), \ a(0) \right]. \tag{15}$$

Upon defining Equation (15), we note that the second integral of its right-hand-side is the value function in the interval $[t, \infty)$. Additionally, at reducing the discount factor ρ, expanding the first order derivative of $v(\cdot, t)$, eliminating the expectation symbol, among other manipulations, we get $\forall \alpha \in \Omega$:

$$\rho v(\alpha, x, a) = \min_{(u(t), w(t)) \in \Gamma(\alpha)} \left[g(\cdot) + [u(t) - d[1 + \beta(u(t))]] \frac{\partial v(\alpha, x, a)}{\partial x} + [ku(t)] \frac{\partial v(\alpha, x, a)}{\partial a} \right.$$

$$\left. + \sum_{\alpha \in \Omega} \lambda_{\alpha\alpha\prime}(\cdot) v(\alpha, x, \varphi(\xi, a))(\alpha) \right] \tag{16}$$

with the following reset function:

$$\varphi(\xi, a) = \begin{cases} 0 & \text{if } \xi(\tau^+) = 1 \text{ and } [\xi(\tau^-) = 4 \text{ and } w(t) = 1] \\ a(\tau^-) & \text{otherwise} \end{cases}. \tag{17}$$

Condition (10) implies that a major maintenance restores the cumulative age to a zero value. Then Equation (17) models the benefit of major maintenance. Further, $\frac{\partial}{\partial x} v(\cdot)$ and $\frac{\partial}{\partial a} v(\cdot)$ refer to the partial derivatives of the value function $v(\alpha, x, a)$. The importance of Expression (16) relies on the fact that it is the fundamental Hamilton-Jacobi-Bellman (HJB) equation, which defines a sufficient condition for an optimum. The procedure to obtain the HJB equations can be consulted in Gershwin [17] and references therein. However, the HJB equations are typically unsurmountable to analytically solved, as noted by Hlioui et al. [18] and there are relatively few exceptions for simple trivial cases. In the next section, we detail the adopted approach to determine the optimal feedback control.

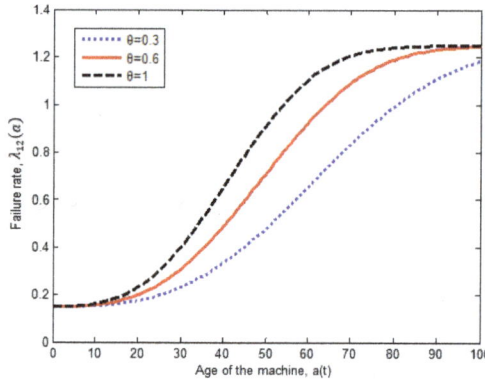

Figure 4. Increasing failure rate.

4. Optimization Method Description

With the aim to determine the optimal joint control policy, i.e., the optimal value of the production rate, and the optimal repair/major maintenance scheduling, we use a numerical technique called Kushner's approach to solve the HJB Equation (16). Such technique was proposed by Kushner and Dupuis [19] and Gharbi et al. [20], and the idea behind this procedure is to approximate the value function $v(\alpha, x, a)$ by a discrete function $v^h(\alpha, x, a)$, and the first-order partial derivatives of the value function $\partial v(\cdot)/\partial x$ and $\partial v(\cdot)/\partial a$ are approximated by:

$$\frac{\partial v}{\partial x}(\alpha, x, a) = \begin{cases} \frac{1}{h_x}(v^h(\alpha, x + h_x, a) - v^h(\alpha, x, a)) & if \ \dot{x} \geq 0 \\ \frac{1}{h_x}(v^h(\alpha, x, a) - v^h(\alpha, x - h_x, a)) & otherwise \end{cases} \tag{18}$$

and

$$\frac{\partial v}{\partial a}(\alpha, x, a) = \frac{1}{h_a}(v^h(\alpha, x, a + h_a) - v^h(\alpha, x, a)) \tag{19}$$

where h_x and h_a define the length of the finite difference interval of the state variables (x, a). The Kushner's technique is useful because it converts the continuous minimization problem to a discrete-time, discrete-state decision process, for further details about this technique Dehayem-Nodem et al. [21] can be consulted. Without loss of generality, we study the case of a manufacturing system with three defective rates depending on its productivity. Such rates are defined as follows:

$$\beta(u(t)) = \begin{cases} b_1 & if \ u(t) < u_1 \\ b_2 & if \ u(t) \in (u_1, u_2] \\ b_3 & if \ u(t) \in (u_2, u_{max}] \end{cases} \tag{20}$$

where $b_3 \geq b_2 \geq b_1$, and $0 \leq u_1 \leq u_2 \leq u_{max}$. Thus, due to the consideration of production-dependent-defectives, we obtain the following approximated valued functions for the operational mode:

At mode 1:

$$v^h(1, x, a) = \begin{cases} v_1^h(1, x, a) & if\ u(t) < u_1 \\ v_2^h(1, x, a) & if\ u(t) \in (u_1, u_2] \\ v_3^h(1, x, a) & if\ u(t) \in (u_2, u_{max}] \end{cases} . \tag{21}$$

with:

$$v_1^h(1, x, a) =$$

$$\min_{\substack{(u(t), w(t)) \in \Gamma(\alpha) \\ u(t) \in [0, u_1)}} \left[\frac{g(\cdot) + \frac{u(t) - d(1 + \beta(u_1))}{h_x} \left[\begin{array}{c} v^h(1, x + h_x, a) \cdot Ind\{\dot{x} \geq 0\} + \\ v^h(1, x - h_x, a) \cdot Ind\{\dot{x} < 0\} \end{array} \right] + \frac{\dot{a}}{h_a}[v^h(1, x, a + h_a)] + \lambda_{12}(a)v^h(2, x, a)}{\left(\rho + \frac{u(t) - d(1 + \beta(u_1))}{h_x} + \frac{\dot{a}}{h_a} + |\lambda_{aa}| \right)} \right]$$

$$v_2^h(1, x, a) =$$

$$\min_{\substack{(u(t), w(t)) \in \Gamma(\alpha) \\ u(t) \in (u_1, u_2]}} \left[\frac{g(\cdot) + \frac{u(t) - d(1 + \beta(u_2))}{h_x} \left[\begin{array}{c} v^h(1, x + h_x, a) \cdot Ind\{\dot{x} \geq 0\} + \\ v^h(1, x - h_x, a) \cdot Ind\{\dot{x} < 0\} \end{array} \right] + \frac{\dot{a}}{h_a}[v^h(1, x, a + h_a)] + \lambda_{12}(a)v^h(2, x, a)}{\left(\rho + \frac{u(t) - d(1 + \beta(u_2))}{h_x} + \frac{\dot{a}}{h_a} + |\lambda_{aa}| \right)} \right]$$

$$v_3^h(1, x, a) =$$

$$\min_{\substack{(u(t), w(t)) \in \Gamma(\alpha) \\ u(t) \in (u_2, u_{max}]}} \left[\frac{g(\cdot) + \frac{u(t) - d(1 + \beta(u_{max}))}{h_x} \left[\begin{array}{c} v^h(1, x + h_x, a) \cdot Ind\{\dot{x} \geq 0\} + \\ v^h(1, x - h_x, a) \cdot Ind\{\dot{x} < 0\} \end{array} \right] + \frac{\dot{a}}{h_a}[v^h(1, x, a + h_a)] + \lambda_{12}(a)v^h(2, x, a)}{\left(\rho + \frac{u(t) - d(1 + \beta(u_{max}))}{h_x} + \frac{\dot{a}}{h_a} + |\lambda_{aa}| \right)} \right] .$$

In essence, for the case of considering three defectives rates, the Kushner's technique defines a total of six HJB Equations. Where three equations are used for the operational mode, and we have three additional equations: one for the failure, minimal repair, and major maintenance modes, respectively. As we can note the number of HJB Equations increases compared to the case of a manufacturing system without the productivity-dependent defectives rate assumption.

5. Simulation and Numerical Results

In this section, we determine the structure of the joint optimal control policy that considers production-quality and maintenance aspects in an integrated model. We solve the discrete event dynamic programming problem (21) through the value iteration procedure, which is detailed in Hajji et al. [22]. In such procedure, the solution of $v^h(\cdot)$ is an approximation that will converge to the solution of $v(\cdot)$ of Equations (14) as $h_x \to 0$ and $h_a \to 0$, with the corresponding boundary conditions defined by the finite grid G_{ax}. The implementation of the approximation technique requires the use of a finite grid G_{ax}, which is defined as follows:

$$G_{ax} = \{(a, x):\quad 0 \leq a \leq 100, -10 \leq x \leq 10\} \tag{22}$$

The limiting probabilities of modes $\xi(t) \in \Omega = \{1, 2, 3, 4\}$, (i.e., π_1, π_2, π_3 and π_4) are computed with the following expressions:

$$\pi \cdot Q(\cdot) = 0 \text{ and } \sum_{i=1}^{\Omega} \pi_i = 1 \tag{23}$$

where $\pi = (\pi_1, \pi_2, \pi_3 \text{ and } \pi_4)$ and $Q(\cdot)$ denotes the transition matrix (4). In order to ensure the validity of the results, the production system must satisfy the following feasibility condition:

$$u_{max} \cdot \pi_1 \geq d \cdot [1 + \beta(u(t))] \tag{24}$$

where π_1 denotes the limiting probability at the operational mode of the machine. With the feasibility condition (24), we ensure that the system will be able to satisfy customer demand even in cases of severe deterioration. We note that condition (24) is satisfied by the parameters presented in Table 1, used in the numerical example.

Table 1. Parameters for the numerical example.

u_{max}	d	ρ	h_x	h_a	u_1	u_2	λ_1	b_1	b_2
10	4	0.9	0.5	1	2.5	5	0.15	0	0.2160
λ_2	λ_{23}	λ_{24}	λ_{31}	λ_{41}	θ	η_0	η_1	b_3	r
1.1	120	120	5	4	0.6	1	3	0.7290	$-15 \times 10^{-6.2}$

Further, we define $k = 1$. The cost parameters for the numerical example are presented in Table 2.

Table 2. Cost parameters for the numerical example.

c^+	c^-	c_r	c_m	c_d	c_{pro}
1	150	5	20	3	0.5

The structure of the obtained joint control policy is discussed as follows:

5.1. Production Policy

We present the obtained production policy $u^*(t)$ in Figure 5a. In such policy the production rate applied in the operational mode is defined in function of the system state that is determined in this case by the stock level and the age of the machine, (x, a). At examining this figure, we can realize that it exists a specific threshold for each production rate. To better interpret the production policy, we present its trace in Figure 5b. From the analysis of this figure, we have the following observations:

i.   The computational domain is divided in four zones, where the production rate is set to 0, u_1, u_2, or u_{max}, respectively. Such zones are delimited by the production thresholds Z_1, Z_2, and Z_3.

ii.   The production rate is decremented gradually (i.e., $u_{max} > u_2 > u_1 > 0$) as the stock level increases and surpasses the production thresholds Z_3, Z_2, and Z_1 with $Z_1 > Z_2 > Z_3$. We note that the minimum production rate u_1 is recommended for high levels of inventory, when the stock level is between the production threshold Z_1 and Z_2. Since given the existence of inventory, the system can manage to satisfy product demand operating at a reduced pace and mainly because it avoids further deterioration. Moreover, we note that as the stock level decreases, the production rate increases, thus the machine operates at its second production rate u_2 when the inventory level is between the thresholds Z_2 and Z_3, with the aim to replace inventory faster. Additionally, we note that the use of the maximum production rate u_{max} is limited, because it deteriorates more rapidly than the machine, and so its use is only recommended in scenarios where the stock level is almost depleted and there is a need to rapidly replace the inventory level and avoid further shortages.

iii. From the obtained results we observe that the production thresholds Z_1, Z_2, and Z_3 increases as the machine ages, this reflects the impact of the deterioration process on the production control rule. For instance at age $a = 0$, the production threshold Z_1 has a value of $Z_1 = 4$, and as the machine deteriorates, at age $a = 100$, such threshold increases to $Z_1 = 7$. The same pattern is observed for thresholds Z_2 and Z_3. The increment is explained by the fact that as the machine deteriorates, it produces more defectives and also it is subject to experience more frequent failures. Thus, more inventory is needed as protection to mitigate shortages.

Our results differ from the classical results of Kenné et al. [23] with only three production rules. In our case, the production policy defines the following control rules:

1. Decrease the production rate to zero, if the current stock level is above the first production threshold Z_1.
2. Set the production rate to its first productivity u_1, when the current stock level is under the first production threshold Z_1.
3. Increase the production rate to its second productivity u_2, when the stock level is below the second production threshold Z_2.
4. Increase the production rate to its maximum rate u_{max}, when the current stock level is under the third production threshold Z_3.

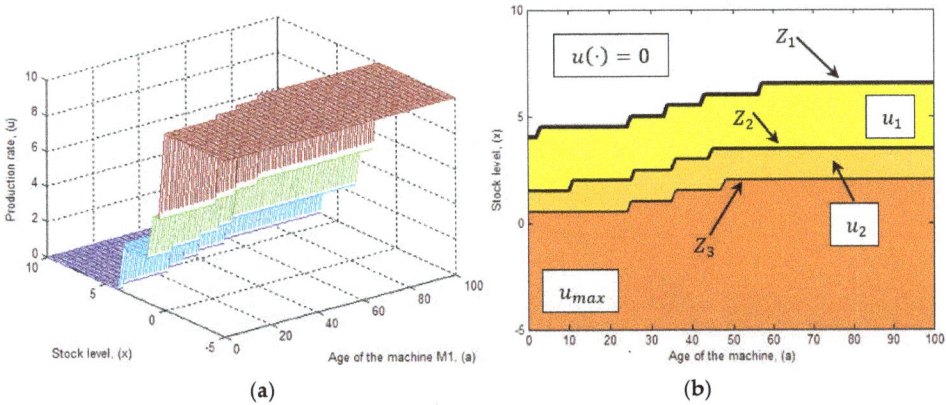

Figure 5. Production policy. (**a**) Production rate; (**b**) Production trace.

In a practical sense, the results of Figure 5 imply a multi-hedging policy form, where the production rate is defined as follows:

$$u(1, x, a)^* = \begin{cases} u_{max} & if \ x(t) < Z_3(\cdot) \\ u_2 & if \ x(t) < Z_2(\cdot) \\ u_1 & if \ x(t) < Z_1(\cdot) \\ 0 & if \ x(t) > Z_1(\cdot) \end{cases} \tag{25}$$

where $Z_1(\cdot) > Z_2(\cdot) > Z_3(\cdot)$ and $0 < u_1 < u_2 < u_{max}$. Equation (25) incorporates the notion that it is advantageous to reduce the production rate from its maximum value to a more profitable level, since production at higher rates accelerates the machine degradation increasing then the total cost, and the economic performance of the unit may be compromised if the effects of such degradation are disregarded.

5.2. Repair/Major Maintenance Switching Policy

The repair/major maintenance switching policy is presented in Figure 6a. Once the machine is at failure we use the decision variable $w(t)$ to properly synchronize the two available maintenance options. The logic behind the maintenance strategy is as follows:

- When $w(t) = 1$, the variable is set to its maximum value, and it denotes the conduction of a major maintenance that completely mitigates the effects of the deterioration process, in this case reducing the failure intensity and the generation of defective units to initial conditions.

- When $w(t) = 0$, the variable is set to its minimum value, indicating that a minimal repair must be performed. This type of maintenance does not restore the machine, since it leaves it the level of deterioration in the same level before the conduction of the repair.

At observing the results of Figure 6a, we note that the computational domain is clearly divided into two zones indicating the type of maintenance recommended. With the aim to facilitate the characterization of the maintenance-switching policy, we present its trace $T_m(\cdot)$ in Figure 6b. This trace serves us to define the following zones:

- Zone M_0: this zone suggests the conduction of a minimal repair, hence the decision variable $w(t)$, is set to its minimum value, $w(t) = 0$.
- Zone M_1 : in this zone the recommendation is to conduct a major maintenance, since given the level of deterioration of the machine is not profitable its operation. Thus $w(t)$ is set to its maximum value, $w(t) = 1$.

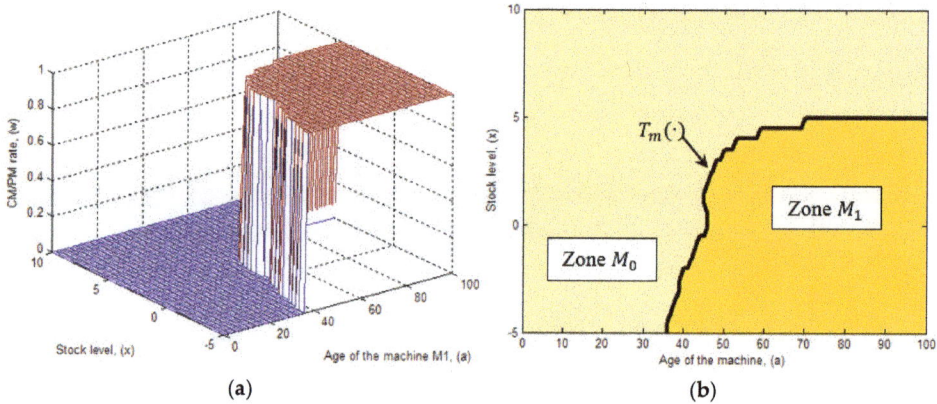

Figure 6. Repair/major maintenance switching policy. (**a**) Repair/major maintenance rate; (**b**) Trace of the repair/major maintenance policy.

Fundamentally, at analyzing the results of Figure 6b, it is evident that the level of deterioration of the machine is the key parameter to determine the maintenance strategy. In particular, we note that when the level of deterioration of the production unit is moderate, (i.e., between age $a = 0$ and $a = 35$ in Figure 6b) then the system opts to recommend a minimal repair and avoid further costs of the expensive major maintenance. However, when the machine reaches higher levels of deterioration (i.e., after age $a = 35$ in Figure 6b) it is preferable to conduct a major maintenance than to experience the effects of more frequent failures and the increase of defective units. Thus, in our results, the repair/major maintenance activities are triggered according to a machine-deterioration-depended policy with a bang-bang structure, and it is given by the following expression:

$$w(2, x, a)^* = \begin{cases} 1 & if \ (a, x) \in Zone \ M_1 \\ 0 & if \ (a, x) \in Zone \ M_0 \end{cases}. \tag{26}$$

In view of Equation (26), once the machine is at failure, major maintenance must be conducted only when the machine has reached a certain level of deterioration that justifies its expensive cost.

6. Sensitivity and Results Analysis

The operational validity of the model is discussed in this section through the analysis of different manufacturing scenarios consisting in the variation of several cost parameters. The purpose of this sensitivity analysis is to determine if the obtained joint control policy is confirmed and characterized consistently by the control factor $Z_i(\cdot)$ and $T_m(\cdot)$. The sensitivity of the proposed control policy is performed according to the variation of the inventory, backlog, repair, major maintenance, and defectives costs. Additionally, we also discuss the influence of an adjustment parameter of the trend of the failure rate.

6.1. Influence of the Backlog Cost

The effect of the variation of the backlog cost on the production policy is presented in Figure 7, where we illustrate the production traces for two cost values, $c^- = 50$ and $c^- = 250$. From the obtained results, we realize that at reducing the backlog cost to $c^- = 50$ the production thresholds are less extended on the computational domain. Then at increasing the backlog cost to $c^- = 250$, the production thresholds increase as protections against backlog, since shortages are more penalized. The influence of the deterioration process on the production policy is evident, because the production thresholds increase as the machine ages. Additionally, we note that the machine operates at a reduced pace (i.e., zone u_1 and u_2) when there is some inventory, and the maximum production rate is devoted just in case of backlog or to maintain some protection stock when the machine reaches higher levels of deterioration.

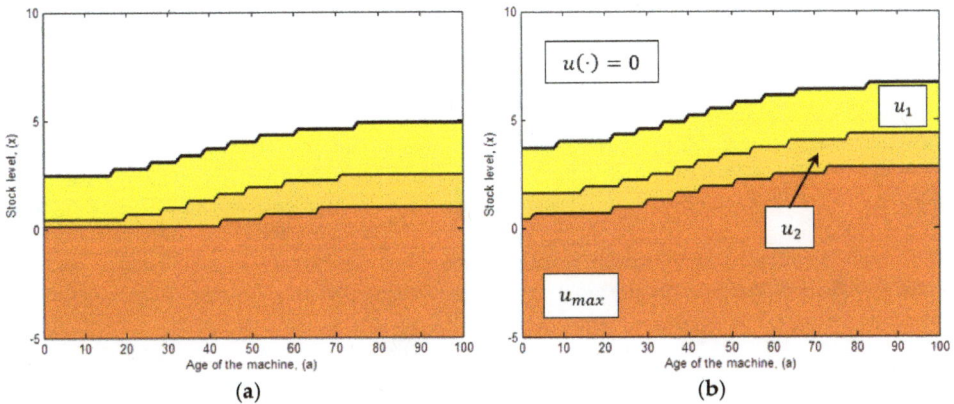

Figure 7. Sensitivity of the backlog cost on the production policy. (a) $c^- = 50$; (b) $c^- = 250$.

Regarding the results of Figure 8, it is apparent that the backlog cost also influences the repair/major maintenance switching policy. For this cost parameter, we compared three cases, $c^- = 50$, 100 and 200. In analyzing such scenarios, we notice that when the backlog cost decreases to $c^- = 50$, major maintenance is less recommended, and so more minimal repairs are performed. If the backlog cost increases to $c^- = 100$, the major maintenance zone grows. Moreover, when the backlog cost is set to $c^- = 200$, even more major maintenance is recommended. This increment is explained because at increasing the backlog cost, the effects of deterioration on shortages and defectives are more penalized. Hence, maintenance options that serve to mitigate effectively the effects of such deterioration, (as major maintenance), are more recommended. With respect to the inventory cost, we notice that is has the opposite effects that the backlog cost.

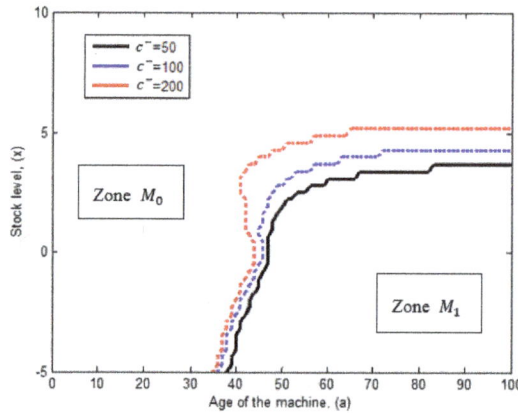

Figure 8. Sensitivity of the backlog cost on the repair/major maintenance policy.

6.2. Influence of the Major Maintenance Cost

In discussing the sensitivity of the major maintenance cost, we examine four scenarios for the cost values $c_m = 20$, 30, 40 and 60. The obtained results are presented in Figure 9. When we set the major maintenance cost to a low value such as $c_m = 20$ and 30, the maintenance policy indicates that the major maintenance zone covers a greater surface on the plane (x, a). Nevertheless, if we increase the major maintenance cost to $c_m = 40$ and 60, minimal repairs are more recommended, reducing considerably the conduction of major maintenance. The observed pattern in the maintenance policy implies the fact that as beneficial and productive as a major maintenance can be, at increasing the c_m cost, the machine must reach higher levels of deterioration to justify the expensive cost of such type of maintenance. Regarding the production policy, we observe that the major maintenance cost has no influence on this policy. Furthermore, that the minimal repair cost c_r has the inverse effects of the major maintenance cost.

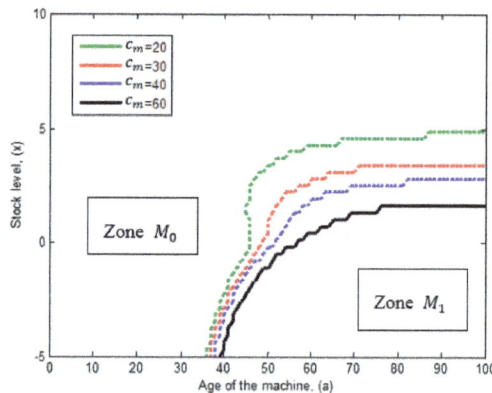

Figure 9. Sensitivity of the major maintenance cost on the repair/major maintenance policy.

6.3. Influence of the Defectives Cost

We proceed with the analysis of the sensitivity of the defectives cost. In Figure 10 we analyze the production trace for two different cost values $c_d = 3$ and $c_d = 12$. From the results we highlight the fact that when the defectives cost is low, for instance $c_d = 3$, there is more liberty to operate the machine at increased paces such as u_2 and u_{max}, regardless that in such rates the machine deteriorates faster, thus producing even more defectives. Moreover, we observe that when we increase the defectives cost to $c_d = 12$, the production threshold Z_1 increases, favoring more the use of the machine at a reduced rate u_1, (since at this mode the machine decreases considerably its deterioration pace, hence it generates less defectives). Additionally, it is evident that the greater the value of the defectives cost, the less extensive is the use of the machine at faster production rates such as u_2 and u_{max}. Because operating at faster production rates, the machine accelerates its deterioration level, generating more defectives.

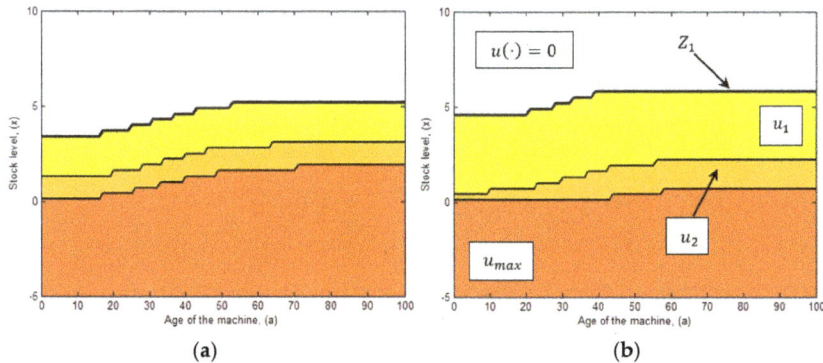

Figure 10. Sensitivity of the defectives cost on the production policy. (**a**) $c_d = 3$; (**b**) $c_d = 12$.

To complement the analysis of the defectives cost, we examine its variation for three cost scenarios, $c_d = 2$, 3 and 4 as illustrated in Figure 11. We begin the discussion when the defectives cost is set to a low value of $c_d = 2$, in this context, the major maintenance activity is limited to the smallest area in the analysis. If the defectives cost increases to $c_d = 3$, the repair/major maintenance trace varies and the area for the major maintenance increases. At increasing the defectives cost to $c_d = 4$, minimal repair is less recommended, whereas the major maintenance zone increases even more. These results amount to the observation that the repair/major maintenance policy is highly sensitive to the defectives cost, since at increasing c_d, more major maintenance is conducted with the aim to restore the machine faster and reduce considerably the generation of defectives and at the same time maintain a reasonable total cost.

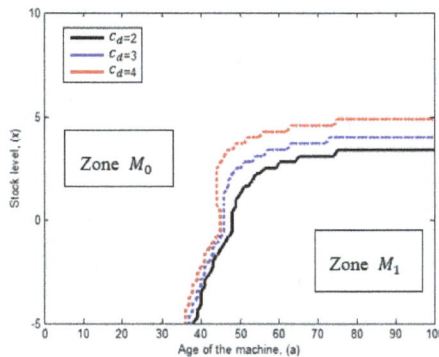

Figure 11. Sensitivity of the defectives cost on the repair/major maintenance policy.

6.4. Influence of the Production Cost

The variation of the production cost c_{pro} indicates that it has a strong effect on the production policy, as can be seen in Figure 12. For this cost we analyze two different cost scenarios with values $c_{pro} = 0.5$ and 5. If the production cost is reduced to $c_{pro} = 0.5$, the production thresholds (Z_1, Z_2 and Z_3) increase, because it is less expensive to operate the machine. By increasing the production cost to $c_{pro} = 5$, the production thresholds (Z_1, Z_2 and Z_3) reduce considerably. The reason behind this decrement is that at increasing the production cost, the system reacts by maintaining just the necessary amount of inventory to palliate shortages caused by failures and defectives units.

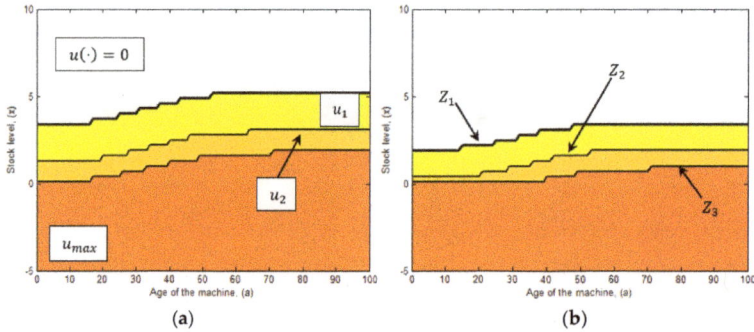

Figure 12. Sensitivity of the production cost on the production policy. (a) $c_{pro} = 0.5$; (b) $c_{pro} = 5$.

The production cost also influences the repair/major maintenance switching policy as observed on Figure 13. We use three different cost values $c_{pro} = 0.5$, 1.25 and 2.5 for the analysis. The results of Figure 13 clearly indicate that major maintenance is less recommended when the production cost is set to a reduced value of $c_{pro} = 0.25$. At increasing the product cost to $c_{pro} = 1.25$, more major maintenance is conducted. Further, the zone for major maintenance expands even more at increasing the production cost to $c_{pro} = 2.5$. To clarify matters, more major maintenance is recommended at increasing the production cost, because the operation of the machine becomes more selective, promoting more maintenance actions that mitigate completely the effects of deterioration c_{pro}.

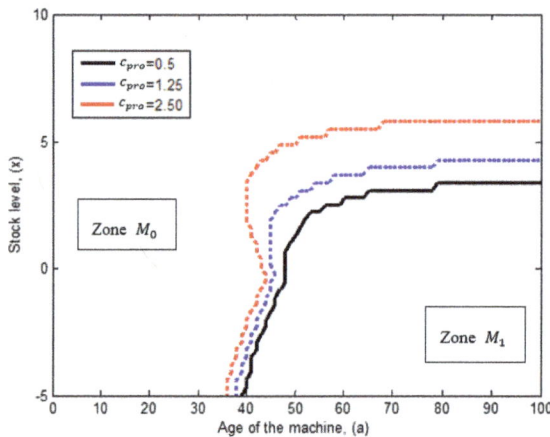

Figure 13. Sensitivity of the production cost on the repair/major maintenance policy.

6.5. Influence of the Pace of the Failure Rate

We complement the sensitivity analysis with the analysis of the effect of the pace of deterioration of the failure rate. For this purpose, we examine the influence of the adjustment parameter θ, which accelerates the failure rate of the machine. For this parameter, we examine two different scenarios with values $\theta = 0.15$ and 1, as illustrated in Figure 14. From the results, we note that whenever the adjustment parameter has a moderate value of $\theta = 0.15$, the production thresholds have a smoother increment as the machine deteriorates. On increasing the adjustment parameter to $\theta = 1$, the production thresholds reach its maximum values more abruptly. Another more pragmatic reason for this pattern is because with the increment of the parameter θ, we are accelerating the pace of deterioration of the machine, and so the production thresholds increase earlier as protections against shortages and defectives.

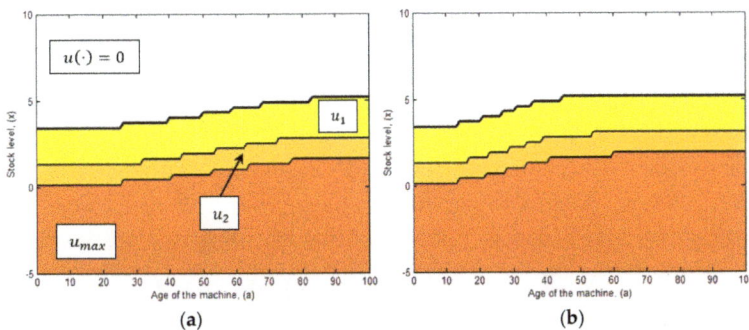

Figure 14. Sensitivity of the pace of the failure rate on the production policy. (**a**) $\theta = 0.15$; (**b**) $\theta = 1$.

A close examination of Figure 15 shows that the repair/major maintenance policy also is significantly affected by the variation of the pace of the failure intensity. To have a better understanding about such influence, we analyze four different scenarios with values = 0.25, 0.50, 0.75 and 1.00. From the maintenance traces presented in Figure 15, we realized that when the adjustment parameter is low, for instances $\theta = 0.25$ and 0.50 it implies that the system experiences less frequent failures, and so major maintenance is less recommended. There are more frequent failures when the parameter is set to a higher value such as $\theta = 0.75$ and 1, where more major maintenance is conducted. From this pattern, we can draw the inference that with higher values of the parameter θ, we accelerate the pace of deterioration of the machine, and this promotes the conduction of more major maintenance to rapidly mitigate the effects of deterioration.

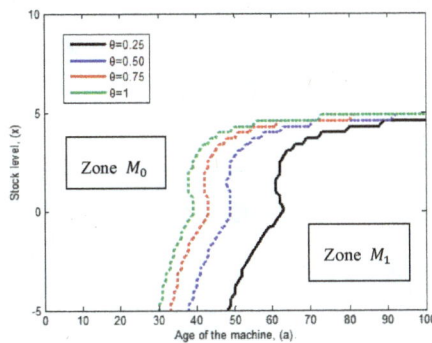

Figure 15. Sensitivity of the pace of the failure rate on the repair/major maintenance policy.

Throughout the discussion of the sensitivity analysis, we observe that the obtained joint control policy is significantly influenced by the deterioration of the machine. Furthermore, this sensitivity analysis allows us to state that the structure of the proposed control policy is consistent, and that such policy is well characterized by the control parameters $Z_i(\cdot)$ and $T_m(\cdot)$. At considering simultaneously in our integrated model, production planning and the repair/switching strategies, we seek to operate more efficiently the manufacturing system.

7. Comparative Study

The objective of this section is to compare the obtained joint control policy in order to illustrate the advantage of applying our approach in practice and highlight the economic benefit that decision-makers can obtain at implementing our joint strategy. We should note that the joint production and maintenance policies proposed in this paper have not been studied under the same assumptions of deterioration and operation-dependent defectives in the literature yet. However, we have managed to compare the total cost incurred from our joint control policy with the total cost reported from other policies based on assumptions common in the literature that does not takes into account the effects of deterioration on the control parameters. The policies considered in the comparison are described as follows:

- Policy-I: defines the production scenario (basic case) of the joint control policy obtained in Section 5, where the production and maintenance policies are determined simultaneously in an integrated model. The particularity of Policy-I is that the production threshold is dynamic, it is adjusted progressively in function of the level of deterioration of the machine, and major maintenance can be once the machine has reached a certain level of deterioration that justifies the cost of such type of maintenance. Thus, the conduction of major maintenance has a feedback on the level of deterioration of the machine.

- Policy-II: this policy is derived from the previous policy with the difference that the production threshold is not adjusted progressively, it remains constant at a given level for all the considered time interval and it does not evolve in function of the deterioration of the machine. Nevertheless, the conduction of major maintenance can be conducted based on a feedback with the deterioration of the machine as in Policy-I.

- Policy-III: the production threshold is dynamic and it is adjusted in function of the deterioration of the machine as in Policy I. However, major maintenance is conducted only when the machine has reached its maximum age limit.

Table 3 summarizes the different production scenarios considered in the comparison, it presents the total cost incurred when the level of deterioration of the system recommends the performance of a major maintenance and also such table presents the observed differences in the total cost. The results presented in Table 3, were obtained with the same data parameters shown in Tables 1 and 2.

Table 3. Cost difference of the comparative study.

Scenario	Production Threshold Z	Major Maintenance w	Optimal Total Cost	Cost Difference Δ (%)
Policy-I (basic case)	dynamic	feedback with deterioration	83.16	-
Policy-II	constant	feedback with deterioration	97.96	17.79%
Policy-III	dynamic	conducted at the age limit	113.40	36.36%

The interpretation of the obtained results of Table 3 implies that Policy-II reported a cost 17.79% higher than our joint control Policy-I, because Policy-II fails to consider the strong effect of the deterioration process on the production control rule. Maintaining a constant production threshold is not profitable in Policy-II, because it is evident that as the machine deteriorates it produces more defectives and also it is subject to more frequent failures. Thus, the production threshold must be progressively increased as a countermeasure to avoid shortages and ensure demand satisfaction with flawless units, as in Policy-I.

With respect to the total cost obtained by Policy-III, we noted that it reported a cost 36.36% more expensive than the cost of Policy-I. The observed cost difference is given mainly because the production and maintenance strategies in Policy-III are not determined simultaneously. In Policy-III, the production threshold are first determined taking into account quality and deterioration levels but with the disadvantage to disregard any consideration of the maintenance strategy. Thus, the obtained results indicate that it is not optimal to separate production and maintenance decisions because the delay of major maintenance increases the effects of deterioration in particular the defectives cost. It is evident that the obtained results support the need of an integrated model such as our Policy-I, that incorporates the strong inter-relationship between the key functions of production-quality and maintenance.

8. Managerial Implication

The obtained joint control policy of this paper is a derivation of the Hedging Point Policy that has been successfully implemented in real production systems as in the Boeing Flap Support Business Unit to control production strategies, as indicated in Gershwin [24]. Other implementation in a real production system, is reported by Dror et al. [25] to coordinate production and subcontracting strategies for a chemical company. In our case, the implementation of the obtained control policy in a real case needs complete information about the state variables of the production system, namely the inventory level and the age of the machine (x, a). The benefit of the implementation of our control policy for the decision maker is that the policy permits to operate the production system more smoothly and predictably at scheduling properly production and maintenances rates. Moreover, our policy has the advantage to endure the effects of the set of disruptions encountered in real production (such as shortages, defectives, increasing failures, imperfect repairs, and deterioration).

The implementation of the obtained control policy is facilitated with the use of the implementation chart presented in Figure 16. This chart defines convenient actions that should be scheduled when the machines is operational and when it is at failure. In the implementation chart the control parameters $Z_1(\cdot)$, $Z_2(\cdot)$, $Z_3(\cdot)$ and $T_m(\cdot)$ must be updated regularly as the machine deteriorates.

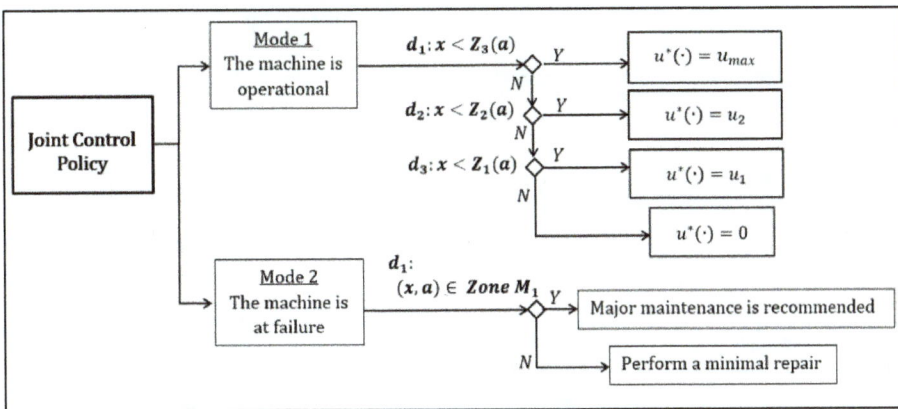

Figure 16. Implementation chart.

As an illustration of how to apply our joint control policy in practice, we determine the production and maintenance control rates for five different points located on the grid (x, a) as presented in Figure 17, for the results obtained in the numerical example of Section 5.

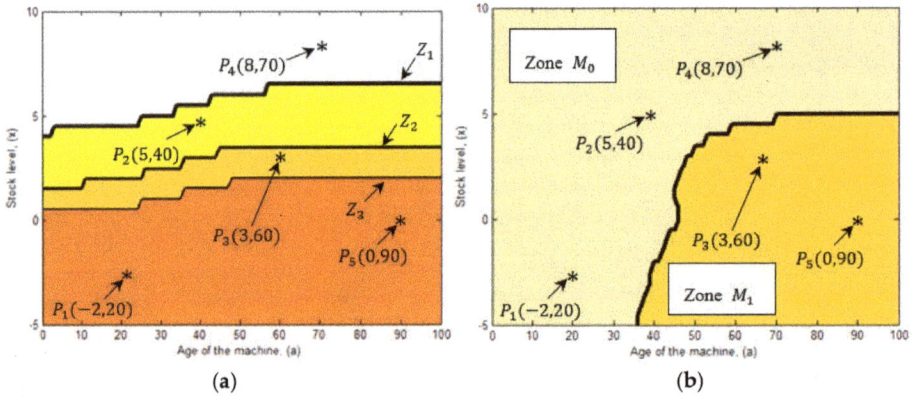

Figure 17. Implementation chart. (**a**) Trace of the production policy; (**b**) Trace of the repair/major maintenance policy.

The value of the control parameters are presented in Table 2 for the selected points (P_1 to P_5) of Figure 17. The value of such control parameters are determined by monitoring the state of the system (x, a) by the implementation chart presented in Figure 16.

The logic of the policy is straightforward, for instance when the machine is operational in point P_3, the stock level is $x = 3$ and the age of the machine is $a = 60$, and so the machine must operate at $u_2 = 5$ because the stock level is less than the productions threshold Z_2 (i.e., $(x = 3) < Z_2$) as indicated in Equation (25). Furthermore, once the machine is at failure, in point P_3 major maintenance is recommended, thus $w = 1$, since the state of the system is inside Zone M_1, where the machine is so deteriorated that the cost of a major maintenance is justified. From the implementation of Table 4, we notice that obtained joint control policy is practical for factory control, because of its ease to be implemented.

Table 4. Implementation of the control policy.

Point	(x, a)	Z	$u^*(\cdot)$	$w^*(\cdot)$
P_1	$(-2, 20)$	Z_3	$u_{max} = 10$	0
P_2	$(5, 40)$	Z_1	$u_1 = 2.5$	0
P_3	$(3, 60)$	Z_2	$u_2 = 5$	1
P_4	$(8, 70)$	Z_1	0	0
P_5	$(0, 90)$	Z_3	$u_{max} = 10$	1

9. Conclusions

The number of scientific publications in the new field that analyses the strong interaction between production logistics, quality, and maintenance design is increasing steadily, reflecting the major relevance of this subject. However, most of the literature is based on systems that optimize either the production-maintenance or the production-quality relationships, leading to sub-optimal solutions. By contrast, this paper investigates the problem of determining the optimal production and repair/major maintenance switching maintenance strategies in the context of reliability and quality deterioration. In the manufacturing system under study, the rate of defectives depends on the production pace of the machine, defining operation-dependent defectives. We developed a stochastic optimization model where the joint production and maintenance control policies were determined by the resolution of the Hamilton-Jacobi-Bellman equations. From the obtained results it has been found when the rate of defectives depends on the production pace where the production policy defines a multi-hedging point policy with several production thresholds. Additionally, the results shows that in order to reduce the total incurred cost, it is beneficial to progressively decrease the production pace of the unit from its maximal value to inferior rates when there are no backlog in the system and there is a positive level of inventory. The results obtained promote major maintenance activities when the production unit reaches a level of deterioration that justifies the cost of an expensive maintenance. We illustrated and validated our proposed approach through a numerical example and an extensive sensitivity analysis. We discussed managerial implications for the decision maker at implementing our approach in real production systems. To further validate the model, a comparative study has been performed, where we noted that our approach represents a valuable alternative for controlling modern production units. Since we obtained cost economies of around 36%, at comparing our joint control policy with other strategies based on common assumptions of the literature that disregard the major interactions among production, quality, and maintenance. As a subject of future research, imperfect maintenance strategies could be integrated to the developed model in order to reduce the total cost and improve the profitability of the company.

Acknowledgments: The authors would like to acknowledge the financial support of PRODEP of Mexico.

Author Contributions: Héctor Rivera-Gómez and Eva Selene Hernández-Gress conceived and designed the mathematical model, Oscar Montaño-Arango and José Ramón Corona-Armenta developed the numerical approach; Irving Barragán-Vite performed the sensitivity analysis and Jaime Garnica-González wrote the paper.

Conflicts of Interest: The authors declare no conflict of interest.

References

1. Liberopoulos, G.; Papadopoulos, C.T.; Tan, B.; Smith, J.M.; Gershwin, S.B. *Stochastic Modeling of Manufacturing Systems*, 1st ed.; Springer: Berlin, Germany, 2006; pp. 3–363. ISBN 978-3-540-26579-5.
2. Colledani, M.; Tolio, T. Integrated analysis of quality and production logistics performance in manufacturing lines. *Int. J. Prod. Res.* **2011**, *49*, 485–518. [CrossRef]
3. Yedes, Y.; Chelbi, A.; Rezg, N. Quasi-optimal integrated production, inventory and maintenance policies for a single-vendor single-buyer system with imperfect production process. *J. Intell. Manuf.* **2012**, *24*, 1245–1256. [CrossRef]
4. Rivera-Gómez, H.; Gharbi, A.; Kenné, J.P.; Montaño-Arango, O.; Hernández-Gress, E.S. Production control problem integrating overhaul and subcontracting strategies for a quality deteriorating manufacturing system. *Int. J. Prod. Econ.* **2016**, *171*, 134–150. [CrossRef]
5. Mhada, F.Z.; Ouzineb, M.; Pellerin, R.; El-Hallaoui, I. Multilevel hybrid method for optimal buffer sizing and inspection stations positioning. *SpringerPlus* **2016**, *5*, 2045. [CrossRef] [PubMed]
6. Bouslah, B.; Gharbi, A.; Pellerin, R. Joint economic design of production, continuous sampling inspection and preventive maintenance of a deteriorating production system. *Int. J. Prod. Econ.* **2016**, *173*, 184–198. [CrossRef]
7. Midfal, I.; Hajej, Z.; Dellagi, S. Production/Maintenance Control of Multiple-Product Manufacturing System. In Proceedings of the Second World Conference on Complex Systems, Agadir, Morocco, 10–12 Novenber 2014. [CrossRef]
8. Khatab, A.; Ait-Kadi, D.; Rezg, N. Availability optimization for stochastic degrading systems under imperfect preventive maintenance. *Int. J. Prod. Res.* **2014**, *52*, 4132–4141. [CrossRef]
9. Hajej, Z.; Turki, S.; Rezg, N. Modelling and analysis for sequentially optimizing production, maintenance and delivery activities taking into account product returns. *Int. J. Prod. Res.* **2015**, *53*, 4694–4719. [CrossRef]
10. Askri, T.; Hajej, Z.; Rezg, N. Jointly production and correlated maintenance optimization for parallel leased machines. *Appl. Sci.* **2017**, *7*, 461. [CrossRef]
11. Martinelli, F. Manufacturing systems with a production dependent failure rate: Structure of optimality. *IEEE Trans. Autom. Control* **2010**, *55*, 2401–2406. [CrossRef]
12. Dahane, M.; Rezg, N.; Chelbi, A. Optimal production plan for a multi-products manufacturing system with production rate dependent failure rate. *Int. J. Prod. Res.* **2012**, *50*, 3517–3528. [CrossRef]

13. Haoues, M.; Dahane, M.; Mouss, K.N.; Rezg, N. Production planning in integrating maintenance context for multi-period multi-product failure-prone single-machine. In Proceedings of the 18th IEEE Conference on Emerging Technologies & Factory Automation, Cagliari, Italy, 10–13 September 2013. [CrossRef]

14. Kouedeu, A.F.; Kenné, J.P.; Dejax, P.; Songmene, V.; Polotski, V. Production and maintenance planning for a failure-prone deteriorating manufacturing system: A hierarchical control approach. *Int. J. Adv. Manuf. Technol.* **2015**, *76*, 1607–1619. [CrossRef]

15. Dong-Ping, S. *Optimal Control and Optimization of Stochastic Supply Chain Systems*, 1st ed.; Springer: London, UK, 2013; pp. 2–272. ISBN 978-1-4471-4724-4.

16. Love, C.E.; Zhang, Z.G.; Zitron, M.A.; Guo, R. A discrete semi-Markov decision model to determine the optimal repair/replacement policy under general repairs. *Eur. J. Oper. Res.* **2000**, *125*, 398–409. [CrossRef]

17. Gershwin, S.B. *Manufacturing System Engineering*, 1st ed.; PTR Prentice Hall: Englewood Cliffs, NJ, USA, 1994; pp. 1–501. ISBN 013560608X.

18. Hlioui, R.; Gharbi, A.; Hajji, A. Replenishment, production and quality control strategies in three-stage supply chain. *Int. J. Prod. Econ.* **2015**, *166*, 90–102. [CrossRef]

19. Kushner, H.J.; Dupuis, P.G. *Numerical Methods for Stochastic Control Problems in Continuous Time*, 2nd ed.; Springer: New York, NY, USA, 2001; pp. 2–475. ISBN 978-0-387-95139-3.

20. Gharbi, A.; Hajji, A.; Dhouib, K. Production rate control of an unreliable manufacturing cell with adjustable capacity. *Int. J. Prod. Res.* **2011**, *49*, 6539–6557. [CrossRef]

21. Dehayem-Nodem, F.I.; Kenné, J.P.; Gharbi, A. Simultaneous control of production, repair/replacement and preventive maintenance of deteriorating manufacturing systems. *Int. J. Prod. Econ.* **2009**, *134*, 271–282. [CrossRef]

22. Hajji, A.; Gharbi, A.; Kenné, J.P.; Pellerin, R. Production control and replenishment strategy with multiple suppliers. *Eur. J. Oper. Res.* **2011**, *208*, 67–74. [CrossRef]

23. Kenné, J.P.; Dejax, P.; Gharbi, A. Production planning of a hybrid manufacturing-remanufacturing system under uncertainty within a closed-loop supply chain. *Int. J. Prod. Econ.* **2012**, *135*, 81–93. [CrossRef]

24. Gershwin, S.B. Design and operation of manufacturing systems: The control point policy. *IIE Trans.* **2000**, *32*, 891–906. [CrossRef]

25. Dror, M.; Kenneth, R.S.; Candace, A.Y. Deux chemicals inc. goes just-in time. *Interfaces* **2009**, *39*, 503–515. [CrossRef]

applied
sciences

MDPI

Article

Kernel-Density-Based Particle Defect Management for Semiconductor Manufacturing Facilities

Seung Hwan Park †, Sehoon Kim † and Jun-Geol Baek *

Department of Industrial Management Engineering, Korea University, Seoul 02841, Korea;
udongpang@korea.ac.kr (S.H.P.); stanley.kimm@gmail.com (S.K.)
* Correspondence: jungeol@korea.ac.kr; Tel.: +82-2-3290-3396
† Those authors contributed equally to this work.

Received: 2 January 2018; Accepted: 28 January 2018; Published: 1 February 2018

Abstract: In a semiconductor manufacturing process, defect cause analysis is a challenging task because the process includes consecutive fabrication phases involving numerous facilities. Recently, in accordance with the shrinking chip pitches, fabrication (FAB) processes require advanced facilities and designs for manufacturing microcircuits. However, the sizes of the particle defects remain constant, in spite of the increasing modernization of the facilities. Consequently, this increases the particle defect ratio. Therefore, this study proposes a particle defect management method for the reduction of the defect ratio. The proposed method provides a kernel-density-based particle map that can overcome the limitations of the conventional method. The method consists of two phases. The first phase is the acquisition of cumulative coordinates of the defect locations on the wafer using the FAB database. Subsequently, this cumulative data is used to generate a particle defect map based on the estimation of kernel density; this map establishes the advanced monitoring statistics. In order to validate this method, we conduct an experiment for comparison with the previous industrial method.

Keywords: kernel density estimation; particle defect management; particle map; semiconductor manufacturing process

1. Introduction

The worldwide semiconductor market is valued at 333 billion dollars [1]. As the semiconductor industry accounts for 10–15% of the total exports of the Republic of Korea, it has a significant influence on the market economy. Owing to the recent proliferation of electronic devices such as mobile phones and tablet PCs (Personal Computers), market competition is becoming increasingly fierce [2]. In the future, as the demand for the internet of things (IoT) products is expected to increase, most manufacturers endeavor to increase the production through maintenance of facilities [3]. The semiconductor manufacturing process consists of numerous steps. As shown in Figure 1, in the front-end process, the ingot is cut to produce wafers and the circuit is designed. The wafers are subsequently undergone fabrication (FAB), which consists of the following eight processes: oxidation, lithography, etching, strip & clean, ion-implantation, chemical vapor deposition, metal deposition, and chemical mechanical planarization. Oxidation refers to a process whereby a thin and uniform silicon oxide film is formed by the chemical reaction of oxygen or water vapor and the surface of the wafer at a high temperature. Lithography is a process whereby patterns on the wafer is formed by photoresist coating, exposure, and development processes. Etching refers to the process whereby unnecessary portions are selectively removed by using the reactive gas to form a circuit pattern. Strip & clean refers to the process of removing particle contamination generated on the wafer amid other ongoing processes. Ion-implantation is the process whereby characteristics of the electronic device are generated by the implantation of impurities converted to a gas form in the circuit pattern. Chemical vapor deposition is the process whereby the water vapor formed in the particles are formed by the chemical reaction of gases to form an insulating film. Metal deposition interconnects each circuit formed on the surface of a wafer with aluminum and copper wire. Chemical mechanical planarization is a process whereby the oxide film and the metal thin film coated on the wafer are ground and flattened using chemical and physical processes. The subsequent step is the back-end process, in which the probe test, the assembly, and the package tests are conducted sequentially. Thereby, the final product is manufactured. These processes each have their own facilities, and each facility includes several chambers where a wafer must be placed for fabricating. The above-mentioned processes occur within those chambers.

Wafer fabrication for semiconductor devices such as microprocessors, memories, digital signal processors, and consumer electronics applications involves a complex and lengthy process with 30–40 reentrant loops, and the above-mentioned processes (①–⑧), as shown in Figure 1 [4]. This study focuses on these eight processes and the inter-process metrology.

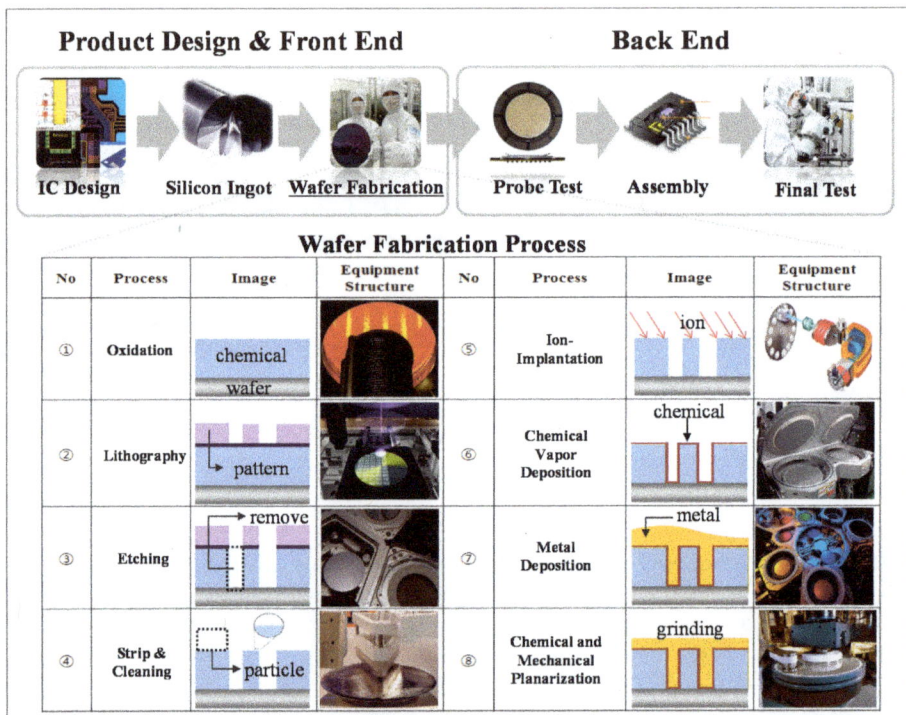

Figure 1. The semiconductor manufacturing process. IC, Integrated Circuit.

Semiconductor FAB includes various physical and chemical treatments and takes six to eight weeks. Therefore, there is a considerable difficulty in identifying causes of failure in the fabrication process, which involves numerous facilities. Figure 2 chronologically shows a pitch change in the lithography process used for fabricating DRAM (Dynamic Random-Access Memory), NAND (Negative-AND gate flash memory), and logic products that have been actively manufactured in recent years. The pitch on the Y-axis illustrates the integration capacity of the semiconductor products and significantly affects product yield, hence the increasing significance of defect management as current FAB processes design and fabricate sophisticated patterns [5].

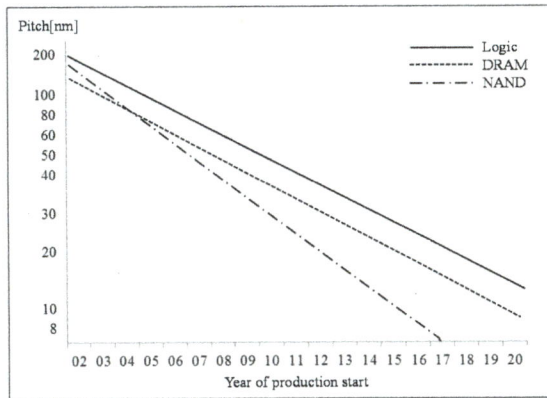

Figure 2. Pitch change in the lithography process (2002–2020). DRAM, Dynamic Random-Access Memory; NAND, Negative-AND gate flash memory.

The defects in a semiconductor manufacturing process are generally distributed as shown in Figure 3. The pie chart on the left shows that the particle defects account for 75% of the total defects. As shown in the pie chart on the right, 75% of the particle defects are caused by process equipment [6].

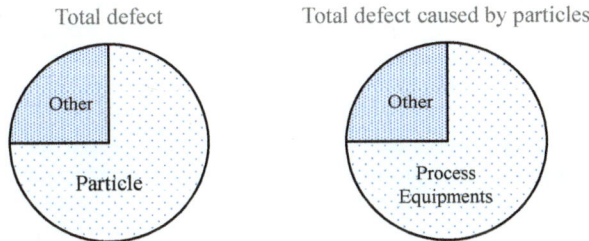

Figure 3. Distribution of defects in a semiconductor manufacturing process.

In the field of semiconductor manufacturing, a particle defect indicates a failure resulting from fine particles, such as dust, that are present on the wafers during FAB. The size of the particle defect owing to the manufacturing environment has remained unchanged, but the FAB process has been improved to develop capabilities for the fabrication of fine patterns. During one inspection, as shown in Figure 4, there was no defect when the pitch was 130 nm, but a pitch of 45 nm resulted in a defect because the size of the particle was greater than the pitch size. Therefore, the structure and stability of the deposited chemical must be carefully controlled, and reduction in contamination in particular becomes increasingly crucial as device sizes shrink [7].

The particles in the chamber of a facility need to be monitored to maintain the chamber's stability. Current manufacturing processes manage particle defects through a simple control chart (c-chart) that monitors the number of particles. As these monitoring charts do not consider the distribution or density of the particles on the wafers, engineers do not know whether the particles are assignable or what the common causes are. Therefore, this study proposes a particle defect management method to overcome the limitation of this monitoring chart.

This paper is organized as follows. Section 2 reviews the conventional particle management method. Section 3 provides the details of the proposed method, and Section 4 presents the experimental results and evaluation. Finally, Section 5 concludes this study.

Pitch	Defect Image	Judgment
130 nm	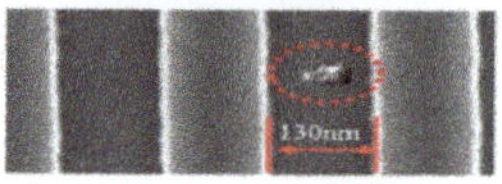	Non-defect
45 nm	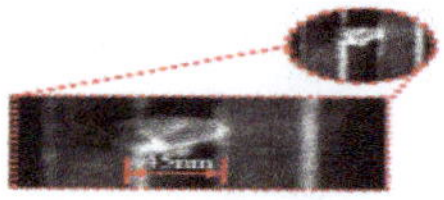	Defect

Figure 4. Change in the particle defect size with respect to the pitch.

2. The Conventional Industrial Method

In an actual semiconductor manufacturing process, conventional particle defect management involves manual action by engineers if, based on the monitoring chart, there is a cause for alarm. The procedure consists of three phases for the detection of the cause of the defect. In the first phase, when an alarm occurs, the insides of the manufacturing equipment are repeatedly cleaned. In the second phase, if the problem persists, the field engineers replace suspicious parts. In the final phase, if the previous replacement does not solve the prevailing problem, predictive maintenance (PM) of the equipment is performed. The second phase, in particular, consumes a substantial amount of time in terms of both decision-making and detection of the root cause, owing to an experience-based decision of the replacement. Therefore, reducing false alarms in the first phase is critical.

Among the aforementioned particle defect management phases, we focus on the first phase, wherein the particle defects are monitored. Particle defects in process equipment are primarily due to equipment aging. Other defects occur due to environmental factors. The counts control chart (c-chart) is commonly used to monitor particle defects. The c-chart is based on the assumption that the distribution of the number of nonconformities is sufficiently well fitted by a Poisson distribution [8]. However, in the actual semiconductor manufacturing process, the particle count does not follow a Poisson distribution [9]. Especially, in our case, random variables are not independent since the particles are cumulated with a time-varying property. Figure 5 shows the metrology equipment gauging the number of particles on the wafer. The number of particles, N, at each coordinate on the wafer is stored.

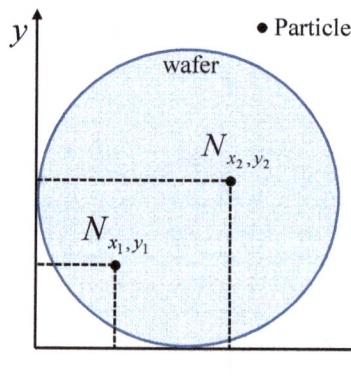

Figure 5. Particle measurements on a test wafer.

Moreover, these particles indicate the status of the equipment, not of the wafer. Therefore, in conventional monitoring statistics, the particle defect (PD) that indicates the number of particles on a test wafer can be denoted as follows:

$$PD = \sum_i N_{x_i, y_i} \tag{1}$$

where x_i and y_i are the i-th particle coordinates, and N_{x_i, y_i} indicates the number of particles at the designated coordinate.

The particle coordinates can be used to generate a wafer map to obtain the distribution and number of particles. Figure 6 depicts four illustrative particle defect maps for four processes. The particle defect maps indicate high-density particles at a particular location on the wafer and the location is correlated with the structure of the process equipment. As indicated by the dotted ellipse in each particle defect map, a high-density area indicates that an ongoing wafer has a potential defect. The particle defect map of the oxidation is also affected by the structure of the equipment. Field engineers have discovered a problem at the center position, where the gas is injected into the furnace, using the particle map. The lithography process shows a higher density at the edges of the wafer. The lithography process includes a photo resist (PR) coating using a high-speed rotation of the spinner equipment in the wafer process. Therefore, the remnants of the chamber affect the density of the wafer edges owing to the characteristics of the high-speed spinning process. Both the etching and strip & clean processes exhibit the highest densities where the dotted ellipse is drawn. Engineers have checked the equipment that analyzes the causes of this phenomenon and discovered a particle resulting from the improper fastening of the O-ring in the chamber. Further, as the strip & clean process includes multiple baths that form bias chemicals at the bottom wafer, the particle defect map indicates high densities of the particles in the bottom area.

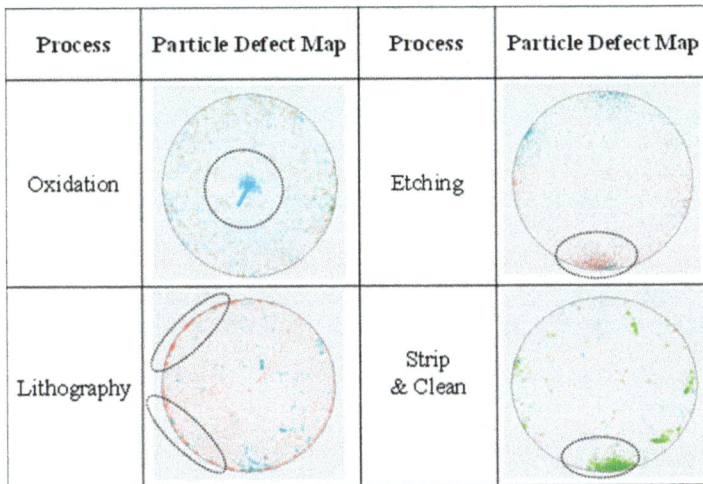

Figure 6. Particle maps of the four processes.

A real domain conducts monitoring based on the control chart in Equations (2)–(4) to monitor the particle counts [9].

$$Center\ line = \overline{PD} \tag{2}$$

$$Upper\ control\ limit = \overline{PD} + CV \times \sqrt{\overline{PD}} \tag{3}$$

$$Lower\ control\ limit = \overline{PD} - CV \times \sqrt{\overline{PD}}. \tag{4}$$

However, *CV* (critical value) is decided through the experience of engineers because the particle defect cannot assume a specific distribution. Moreover, since the particle count is always positive, *Lower control limit* is zero. Therefore, the actual process determines the abnormalities in the facility using the threshold of the particle count, as shown in Figure 7.

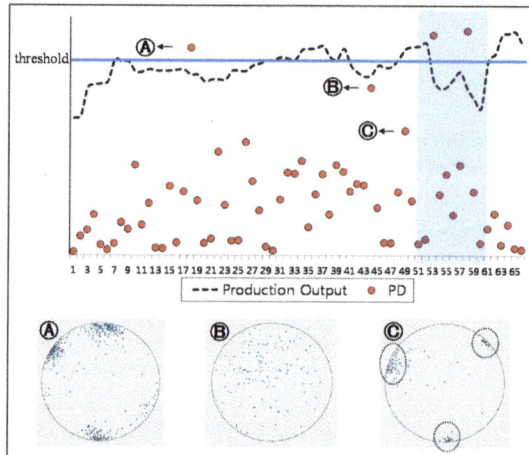

Figure 7. Illustration of the conventional monitoring method. PD, particle defect.

Figure 7 describes both the monitoring chart and the particle maps of the wafers required for observing the particles with specific equipment. In the monitoring chart, the horizontal axis is the wafer index and the vertical axis is the monitoring statistic, *PD*. The dashed-polygonal line indicates production yield and the solid straight line indicates the criterion that distinguishes between a non-defect and a defect. As shown in the chart, Wafer A is beyond the threshold, whereas Wafers B and C are within the threshold. Although both Wafers B and C are within the threshold, they have different distributions, as shown by their respective maps. The three wafer maps depict their respective particle distributions. Wafer A shows high-density particles because *PD* is beyond the threshold. The particles in Wafers B and C are differently distributed, whereas the monitoring chart indicates that *PD* is within the threshold for both wafers. The particles on Wafer B are uniformly distributed, and the particles on Wafer C are concentrated around the edges of the wafer. Consequently, despite a smaller *PD*, the particles concentrated on Wafer C can induce fatal risks, including out-of-threshold wafers or yield reductions, as shown by the shaded rectangular area of the plot. Therefore, since it is important to identify particle distributions, as in Wafer C, this study proposes a new monitoring method considering dense particles.

3. The Proposed Method

The proposed method consists of two stages. The first stage consists of the particle map based on the estimation of kernel density and the cumulative particle data. The second stage proposes a new monitoring statistic, calculated from the probability of the kernel function based on the particle map.

3.1. Particle Map Generation Based on the Kernel Density

As the particle map indicates only the distribution of the particles using the cumulative data, the density of the particles should be considered. Therefore, this method utilizes a function for the estimation of kernel density. Kernel density estimation is a non-parametric method for estimating the probability density of a dataset. As the particle map consists of two dimensions, the X-axis and Y-axis of the wafer, this map is constructed using multivariate kernel density estimation. Considering

$\{x_1, y_1\}, \{x_2, y_2\}, \ldots, \{x_n, y_n\}$ as the sample data of two-dimensional vectors and n as the number of particles, the kernel density estimation function is expressed as [10], where H is the scale coefficient of the kernel function and K is the Gaussian kernel function, which is a symmetric multivariate density.

$$\hat{f}_H(x,y) = \frac{1}{n} \sum_{i=1}^{n} K_H\{(\overline{x}, \overline{y}) - (x_i, y_i)\}. \tag{5}$$

The scale coefficient H of Gaussian kernel function K is estimated using a well-supported rule [11] and is dependent on the performance of particle inspection equipment. H means the estimated sigma. In this study, we selected 99,370 as the value of H. However, this value can vary depending on the resolution of image data from equipment. Consequently, this function calculates the density $\hat{f}_H(x,y)$ of the wafer. We build a three-dimensional kernel-density-based particle map using $\hat{f}_H(x,y)$ in the X-axis and Y-axis. Figure 8a,b illustrate the particle map and the kernel-density-based particle map, respectively. In Figure 8a, the dotted area indicates dense particles, and Figure 8b shows a high density at the corresponding area in Figure 8a. Hence, Figure 8b shows the probability density of the corresponding particles of the map in Figure 8a. This particle map contains dense particles regardless of the total count of the particles.

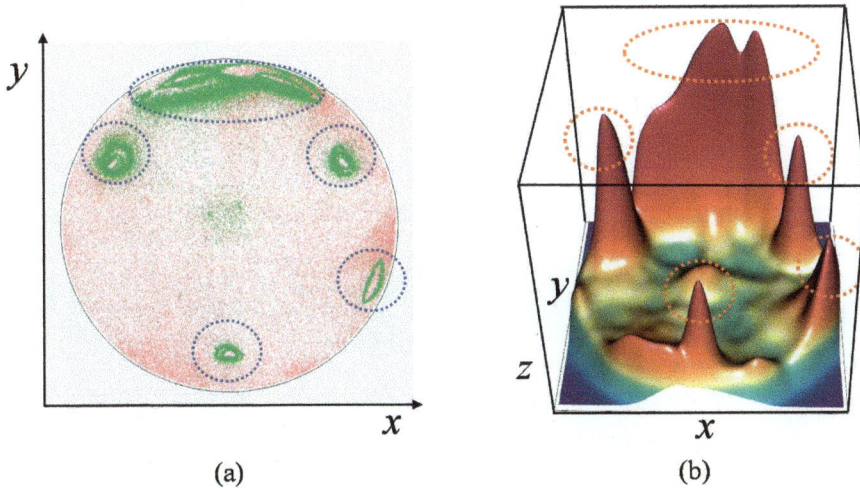

Figure 8. The conventional and new particle maps: (**a**) particle map; (**b**) kernel-density-based particle map.

As these dense particles can cause an abnormal state in the facility, this study devises new monitoring statistics that consider the density of the particles using the kernel-density-based particle map.

3.2. New Monitoring Statistics Using the Proposed Map

As mentioned in the previous section, the conventional method uses *PD* for monitoring. This chapter describes new monitoring statistics that can replace the conventional statistics. The kernel-density-based particle map shows the probabilities at all the points on the wafer, as shown in Figure 8b. While the original particle map shows only the location of the particles, the kernel-density-based particle map includes the density of the particles at a location. In the case of a kernel-density-based particle defect, the expected value E of the particle defect is calculated as

$$E = \frac{\hat{f}_H(x,y)}{\sum_{i=1}^n \hat{f}_H(x_i, y_i)} \times n \qquad (6)$$

where n is the total count of the particles, $\hat{f}_H(x,y)$ denotes the density of the particle map, and $\frac{\hat{f}_H(x)}{\sum_{i=1}^n \hat{f}_H(x_i)}$ is the probability based on the density. The expected value indicates a new particle defect at the location of the particle. Therefore, according to the new monitoring statistics, the kernel-density-based particle defect ($KDPD$) is calculated as

$$KDPD = PD + \sum_i E_{x_i, y_i} \qquad (7)$$

where x_i and y_i are i-th particle coordinates and E_{x_i, y_i} indicates the expected value of each particle at the designated coordinates.

The sum of E_{x_i, y_i} indicates the number of particles considering the densities of the wafer and defines the new monitoring statistics. Thus, the monitoring chart utilizes $KDPD$ to detect potential particle defects. Figure 9 depicts the comparisons between the results of certain PD and $KDPD$. While Wafers A, B, and C exhibit identical PDs, $KDPD$ assumes different values for each wafer. As $KDPD$ is derived using the particle distributions on the map in Figure 9, it indicates a density-based PD. The dense particles, shown at the bottom of the Wafer C map, influence monitoring performance. Thus, $KDPD$ dynamically changes in accordance with the particle distribution.

Wafer	A	B	C
PD	50	50	50
KDPD	54.5	55.5	71
Particle Defect Image	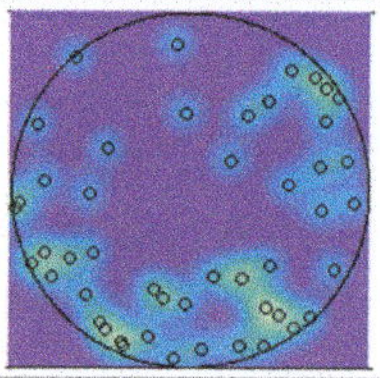	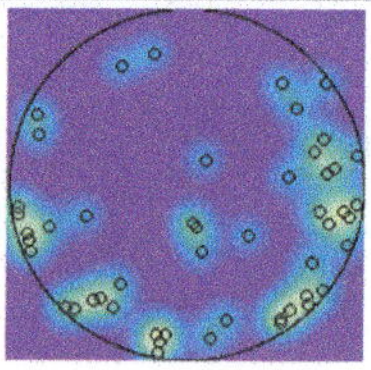	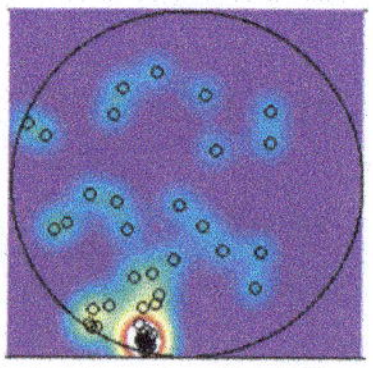

Figure 9. Comparative particle defect (PD) and kernel-density-based particle defect (KDPD).

4. Experimental Results

This section describes an experiment for verifying the proposed method and its results. The particle maps used for this experiment were retrieved from the etching process. As several "killer-particles" originate from the deposition of a layer of by-product on the inside of the plasma-etching chamber, the etching process is sensitive to particles [12]. Therefore, we used particle data from a real etching process in a Korean semiconductor manufacturing company, obtained over a period of five months. The entire dataset consists of 600 wafers, generating 600 particle maps. Among these particle maps, we defined the particle defect types using field knowledge. A dense particle indicates that the facility is abnormal. Therefore, the dataset with dense and even particles was selected. The four defect types are shown in Figure 10. Our dataset consists of 300 normal wafers, 100 wafers of Type 1, 100 wafers of Type 2, 50 wafers of Type 3, and 50 wafers of Type 4. These types were determined by experienced engineers.

As shown in Figure 10, Type 1 shows an illustrative particle map of a small defect. This map includes only one particle cluster, shown on the left side of the map. Type 2 map has two or more small defects. The Type 3 map has only one large-sized defect and the Type 4 map shows the particle distribution on the entire surface of the wafer.

According to the specified defect types, we generated four datasets for the experiment, as shown in Table 1.

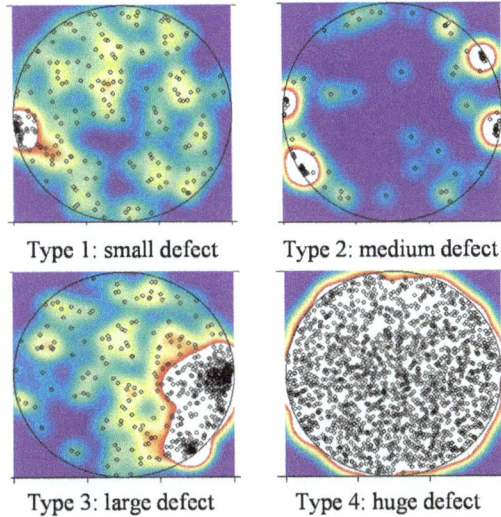

Type 1: small defect   Type 2: medium defect

Type 3: large defect   Type 4: huge defect

Figure 10. Levels of the defects in the etching process.

Table 1. Defect datasets.

Dataset	Combination of Defect Types
Dataset 1	Types 1, 2, 3 and 4
Dataset 2	Types 2, 3 and 4
Dataset 3	Types 3 and 4
Dataset 4	Type 4

Dataset 1 includes non-defect wafer images and wafer images with Defect Types 1, 2, 3 and 4. On the other hand, Dataset 4 consists of non-defect wafers and Type 4 defect wafers only. In order to validate the monitoring performance for these datasets, we evaluated the use of $KDPD$, compared to the use of the conventional monitoring statistics, to identify a defect.

We used the monitoring method illustrated in Figure 7 to classify non-defect and defect particle maps. The threshold in the monitoring chart corresponds to the upper control limit (UCL) in the statistical process control chart. Therefore, we compared the classification performance of the particle defect map by adjusting the decision threshold of the defect/non-defect. Further, we used the receiver operating characteristic (ROC) curve, a graphical plot that illustrates the performance of a binary classification. Moreover, in order to compare these curves, the area under the ROC (AUROC) curve was considered [13]. A broader AUROC can detect a particle defect more accurately. As defect classification is more important than non-defect classification, the ROC curve uses a false negative rate for the X-axis and a true negative rate for the Y-axis. Figure 11 depicts the ROC curves for both PD and KDPD of the four datasets defined in Table 1, and the AUROC examines the monitoring performances.

The ROC curves of Datasets 1, 2 and 3 show that the AUROCs of $KDPD$ are broader than those of PD. Therefore, KDPD outperforms PD. Moreover, the results indicate an increase in the performance gap between PD and KDPD in proportion to the degree of the defects. On the other hand, Dataset 4 demonstrates that the conventional statistics (PD) outperforms KDPD. This dataset includes the Type

4 defect as specified in Table 1 and is referred to as a fatal defect. This can be monitored using either PD or KDPD. Therefore, the proposed KDPD is appropriate for detecting Type 1, 2, and 3 defects. Since the Type 1, 2 and 3 defects indicate potential factors that can result in fatal defectives of Type 4, field engineers can detect critical defects in early stages. Table 2 represents the area under the ROC curves depicted in Figure 11. In Datasets 1, 2 and 3, the area of KDPD is much larger than PD. However, in the case of Dataset 4, the area of KDPD is slightly smaller than PD.

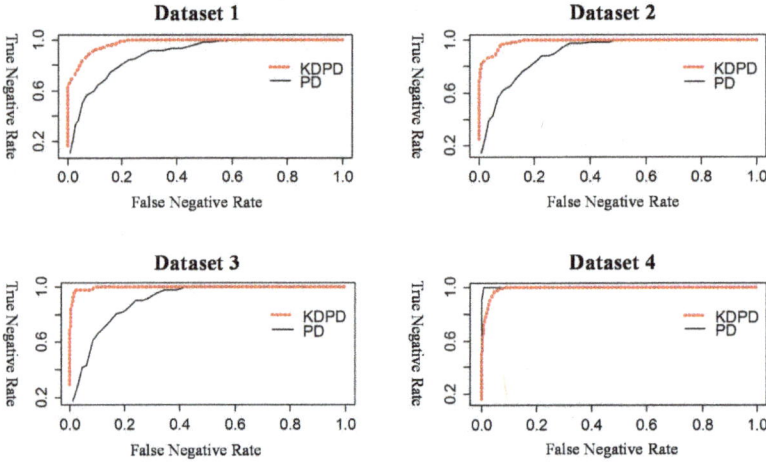

Figure 11. Receiver operating characteristic (ROC) curves of the four datasets including each fault type.

Table 2. Area under the ROC (AUROC) (If AUROC = 1, then the test is perfect. If AUROC = 0.5, then the test is worthless) [1].

Dataset	PD	KDPD
Dataset 1	0.79	0.92
Dataset 2	0.81	0.93
Dataset 3	0.84	0.98
Dataset 4	0.99	0.97

[1] ROC; Receiver operating characteristic; PD, particle defect; KDPD, kernel-density-based particle defect.

5. Conclusions

This study is significant because the proposed method utilizes the real data of the particles in the chamber of a semiconductor process. Novel monitoring statistics that efficiently identify the particles on the wafers in a chamber are proposed in this study. These new statistics are based on multivariate kernel density estimation and indicate the densities of the particles. Thorough management is essential for reducing the failures caused by particles in a semiconductor manufacturing process. This study presents a particle defect management method using a kernel-density-based particle map for improving the conventional method of monitoring the number of particles. In particular, the main contribution of this study is the development of new monitoring statistics. The new statistics that consider the distribution of particles, when applied to an actual process, produce the following three effects. Firstly, the proposed method can reduce the number of unnecessary replacements. Secondly, according to a field internal review, when this study is applied to an actual process, a reduction of approximately 30% in the meantime to repair (MTTR), i.e., the time to find the cause of a failure, is expected. Third, regardless of the number of particles, dense particles increase the possibility of further failures. Therefore, using the proposed particle maps, it is possible to detect defects and predict future failures.

Since the semiconductor manufacturing processes are becoming finer, particle defect management becomes increasingly important. If the proposed kernel-density-based particle map can be applied to an actual process, it is expected to improve both the yield and the quality of the semiconductor product. The size of particle defect has a large influence on yield. The current wafer size is 300 mm, but the wafer size will grow and the defect rate will increase as the process is refined. Therefore, in the future, it is necessary to identify the relationship between the particle map and both the yield and quality of the wafer.

Acknowledgments: This work was supported by the National Research Foundation of Korea (NRF) grant funded by the Korea government (MSIP) (NRF-2016R1A2B4013678). This work was also supported by the BK21 Plus (Big Data in Manufacturing and Logistics Systems, Korea University) and by the Samsung Electronics Co., Ltd.

Author Contributions: Seung Hwan Park and Sehoon Kim contributed equally to this work as co-first authors. They designed and implemented the algorithm to solve the defined industrial problem. Jun-Geol Baek validated the proposed algorithm and guided the research. All authors read and approved the final manuscript.

Conflicts of Interest: The authors declare no conflict of interest.

References

1. Gartner. Available online: https://www.gartner.com/newsroom/id/3282417 (accessed on 27 December 2017).
2. Chik, M.A.; Saidin, M.H.; Hashim, U. Industrial Engineering Roles in Semiconductor Fabrication. In Proceedings of the 11th Asia Pacific Industrial Engineering & Management Systems Conference, Melaka, Malaysia, 7–10 December 2010.
3. Solid State Technology. Available online: http://electroiq.com/blog/2015/09/the-future-of-mems-in-the-iot (accessed on 27 December 2017).
4. Chien, C.F.; Wang, W.C.; Cheng, J.C. Data mining for yield enhancement in semiconductor manufacturing and an empirical study. *Expert Syst. Appl.* **2007**, *33*, 192–198. [CrossRef]
5. Meiling, H. EUVL—Getting ready for volume introduction. In Proceedings of the SEMICON West 2010, San Francisco, CA, USA, 12–16 July 2010.
6. Ahn, K.-H.; Miller, S.J. Particle sampling and element analysis from semiconductor manufacturing equipment. In Proceedings of the International Symposium on Cleanroom Technology and Contamination Control, Seoul, Korea, 19–20 September 1990.
7. Aziz, F.A.; Ahmad, I.H.; Zulkifli, N.; Yusuff, R.M. Particle Reduction at Metal Deposition Process in Wafer Fabrication. In *Manufacturing System*; Aziz, F.A., Ed.; InTech: Rijeka, Croatia, 2012; pp. 1–26, ISBN 978-953-51-0530-5.
8. Montgomery, D.C. *Introduction to Statistical Quality Control*, 7th ed.; Wiley: New York, NY, USA, 2013; pp. 317–335, ISBN 978-1118146811.
9. Kawamura, H.; Nishina, K.; Higashide, M. Control Charts for Particles in the Semiconductor Manufacturing Process. *Econ. Qual. Control* **2008**, *23*, 95–107. [CrossRef]
10. Silverman, B.W. *Density Estimation for Statistics and Data Analysis*; Chapman & Hall: London, UK, 1986; pp. 7–11, ISBN 0-412-24620-1.
11. Venables, W.N.; Ripley, B.D. *Modern Applied Statistics with S-PLUS*, 3rd ed.; Springer: New York, NY, USA, 1999; pp. 113–148, ISBN 978-1-4757-3123-1.
12. Lee, H.-J.; Lin, S.-Y.; Lin, I.-T.; Wei, K.-L.; Chang, S.-Y.; Lian, N.-T.; Yang, T.; Chen, K.-C.; Lu, C.-Y. Post Etch Killer Defect Characterization and Reduction in a Self-aligned Double Patterning Technology. In Proceedings of the 2011 IEEE/SEMI Advanced Semiconductor Manufacturing Conference, Saratoga Springs, NY, USA, 16–18 May 2011.
13. Kang, S.; Cho, S.; An, D.; Rim, J. Using Wafer Map Features to Better Predict Die-Level Failures in Final Test. *IEEE Trans. Semicond. Manuf.* **2015**, *28*, 431–437. [CrossRef]

applied
sciences

MDPI

Article

A Multi-Usable Cloud Service Platform: A Case Study on Improved Development Pace and Efficiency

John Lindström [1,*], Anders Hermanson [2], Fredrik Blomstedt [3] and Petter Kyösti [1]

[1] ProcessIT Innovations R&D Centre, Luleå University of Technology, 971 87 Luleå, Sweden;
    petter.kyosti@ltu.se
[2] Adage AB, C/O BnearIT, Stationsgatan 69, 972 34 Luleå, Sweden; anders.hermanson@adage.se
[3] BnearIT AB, Stationsgatan 69, 972 34 Luleå, Sweden; fredrik.blomstedt@bnearit.se
* Correspondence: john.lindstrom@ltu.se; Tel.: +46-920-491528

Received: 18 December 2017; Accepted: 16 February 2018; Published: 24 February 2018

Abstract: The case study, spanning three contexts, concerns a multi-usable cloud service platform for big data collection and analytics and how the development pace and efficiency of it has been improved by 50–75% by using the Arrowhead framework and changing development processes/practices. Furthermore, additional results captured during the case study are related to technology, competencies and skills, organization, management, infrastructure, and service and support. A conclusion is that when offering a complex offer such as an Industrial Product-Service System, comprising sensors, hardware, communications, software, cloud service platform, etc., it is necessary that the technology, business model, business setup, and organization all go hand in hand during the development and later operation, as all 'components' are required for a successful result.

Keywords: big data; case study; circular economy; data collection and analytics; development; efficiency; improvement; multi-usable cloud service platform; pace

1. Introduction

This paper addresses a micro small and medium-sized enterprise (SME), Adage AB (hereafter the company), in Sweden and its journey of developing and operating a multi-usable cloud service platform for big data collection and analytics. The company is a spin-off from BnearIT AB, an IT-consulting company, and was founded in order to commercialize and productize technology and ideas from BnearIT AB's participation in European research and development projects, customer interactions, and requests. The journey is described as a case study spanning three contexts: construction and prefabrication (monitoring of hardening process for concrete molds), real estate (monitoring of humidity inside of exterior walls), and recycling management (monitoring and optimization of recycling of glass, paper, metals, plastics, etc.). The latter context will be given the most emphasis, as it is the most interesting and developed one.

Many companies have a need to monitor, predict problems or maintenance need, and simulate and optimize production equipment or processes. The equipment, such as vehicles or machines, and buildings and real estate, as well as infrastructure like railways or tunnels, is subject to optimization of time, effort and money spent regarding development, operation, and maintenance. At the same time, more value is expected to be delivered as the surrounding society and technology continuously develop, become more sophisticated, and learn more about what can be expected or demanded. This is a challenge and this paper provides some insight into how a multi-usable cloud service platform, based on the industrial internet and internet-of-things (IoT) paradigms, can efficiently meet this challenge.

The research question in the paper is "how much improvement has been made and how has Adage AB managed to maintain a high development pace and efficiency throughout the development and simultaneous operation of the cloud service platform?" Further, the problem addressed in the paper is how the company has managed to develop applications for three contexts in an efficient and scalable manner. Besides the technical issues, a number of related matters that the company has overcome are also analyzed and discussed.

2. Related Work

The paper spans a number of research areas and the related work below were selected due to that the cloud service platforms are the most central entity and have a large impact on the research question. Further, brought up are also important issues such as security, trust, privacy and legal matters, which all are business critical and support the cloud service platforms. Traditional security may not be possible to reuse in a legacy manner and needed are new security, trust and privacy mechanisms as well as security frameworks. In addition, some benchmarking examples, regarding improvement in development pace and efficiency, are summarized.

2.1. Multi-Usable Cloud Service Platforms for Big Data Collection/Analysis and IoT

As the amount of data that various actors want to process increases, traditional data and analytics are beginning to meet their limits; an emerging solution for large-scale data collection and analytics is cloud analytics. Demirkan and Delen [1] demonstrate a conceptual architecture of service oriented decision-support, which includes data warehouses, online analytic processing, operational systems and end-user components. This is something that many other actors pursue as well, and three main cloud service offerings are distinguishable: Software-as-a-Service (SaaS), Platform-as-a-Service (PaaS) and Infrastructure-as-a-Service (IaaS) [2]. However, there are quite a few more specialized offerings emerging as well, such as Security-as-a-Service and Identity-as-a-Service, complementing the three main ones. Further, Derhamy et al. [3] categorize a number of different cloud service platforms for IoT as: global cloud, peer-to-peer and local cloud. In addition, often "fog" or hybrids of global and internal clouds are positioned in between the global and local cloud paradigms. Regarding the **global cloud**, for instance, some of the largest and most used cloud service/computing platforms, briefly described below, are: Amazon Web Services (AWS), Microsoft Azure, Google Cloud Platform and IBM Bluemix—which are:

- AWS, a subsidiary of Amazon.com, offers a suite of cloud computing services that make up an on-demand computing platform. AWS has more than 70 services, spanning a wide range, including computing, storage, networking, database, analytics, application services, deployment, management, mobile, developer tools and tools for the IoT [4].
- Microsoft Azure is a cloud computing service created by Microsoft for building, deploying, and managing applications and services through a global network of Microsoft-managed data centers. It provides SaaS, PaaS and IaaS, and supports many different programming languages, tools and frameworks, including both Microsoft-specific and third-party software and systems. The offering includes: data analysis, network, storage, databases, IoT, enterprise-integration, development and monitoring [5].
- The Google Cloud Platform is part of a suite of enterprise services from Google Cloud and provides a set of modular cloud-based services with a host of development tools. Their product portfolio includes: computing, storage and databases, networking, big data, machine learning, management tools, developer tools, and identity and security [6].
- IBM Bluemix is a PaaS that supports several services and programming languages including Java, Node.js, Go, PHP, Swift, Python, Ruby Sinatra and Ruby on Rails and can be extended to support other languages such as Scala. It also includes offers such as infrastructure computation, storage, network, mobile, Watson, data analytics and IoT [7].

Thus, there are quite a few multi-usable cloud service platforms. However, many of these do not have an effective and cost-efficient way to integrate and later operate many IoT-devices from a variety of manufacturers on a long-term basis.

Concerning **peer-to-peer**, Derhamy et al. [3] outline, for instance, the platforms and frameworks, etc. from IPSO, Thread, ThingSquare, IzoT, SEP 2.0, AllJoyn and IoTivity, which have approached IoT application development from a device level and support a high level of peer-to-peer operation, a functional solution for home automation as well as device management.

Currently, the **local cloud** is targeted mainly by the Arrowhead Framework [8], which addresses many challenges related to IoT-based automation, and is unique in its support for integration of applications between secure localized clouds. The approach is that IoT devices are abstracted as services in order to enable interoperability between almost all IoT devices. A local cloud based on the Arrowhead Framework provides improvements, compared to global clouds, regarding: realtime data, data and system security, automation system engineering and scalability of automation systems [8]. Thus, the Arrowhead Framework adds a "glue" to various IoT devices, etc. during the development/integration phase and later operation.

Further, with the emergence of service-oriented business processes, Dyche [9] concludes that architecture and infrastructure which include standardized processes for accessing data, the actual platform on which data resides does not matter. By applying a standard set of transformations to the various sources of data and enabling applications to access the data via open standards (e.g., SQL and XML) service requests can access data regardless of system manufacturer. However, regarding the data sources, many data sources such as IoT devices do not conform to standards and require a means to efficiently be integrated.

2.2. Important Issues Necessary to Manage: Security, Trust, Privacy and Legal Matters

Cloud computing presents many promising technological and economic opportunities. Many customers, however, remain reluctant to move their business IT infrastructure completely to the cloud. Birk and Waegner [10] state that one of the main concerns is cloud security and the threat of the unknown. Cloud service providers encourage this perception by not letting their customers see what is behind their virtual curtain. This continues to fuel insecurity on the sides of both providers and customers. Further, Jensen et al. [11] assert that although the economic benefits of moving to a cloud-based platform are clear, since it can reduce capital expenditure (CapEx) and operational expenditure (OpEx), it is still not clear how the technical security issues and the social trust issues should be resolved. Related to, and part of, security and trust is privacy, and access control and trust need to be aware of this [12] to not cause problems as well as how anonymization of data can be used for this [13]. Further, of interest may be also to be able to quantify the level of trust. The issues above need to be sorted out, together with, for instance, the legal requirements imposed by the EU (General Data Protection Regulation, GDPR) and USA, by any organization or company that approaches cloud services with any sensitive information stored, processed or communicated.

2.3. Development Pace and Efficiency—Benchmarking Examples

In response to the need for faster development pace, Fylaktopoulos et al. [14] describe a modular integrated development environment (including a run-time environment) for cloud-based applications which is a platform built to support model-driven development and team collaboration. The platform facilitates rapid development of advanced applications in the cloud and offers a solution for both rapid business development based on predefined components and application development, providing a layered distributed architecture. The result indicates that inexperienced developers are able to create business applications from scratch directly in the cloud and in a significantly smaller timeframe. Further, since there are no binary files, the installation of updates is an easy process with zero downtime. As the platform is also the runtime environment, the deployment of the developed applications is asserted to be instant.

The open source Arrowhead Framework provides architectural definitions of software systems providing the necessary services that enable the implementation of a self-contained local automation cloud. Delsing et al. [15] conclude, based on three cases, that time savings are in the order of approximately 70–80% when using the Arrowhead Framework, compared to legacy technology-based implementations, during the development/integration phase with IoT devices from different vendors.

Potential alternatives to using the Arrowhead Framework and the modular integrated development environment would be to use an existing IoT-platform together with a cloud service platform. However, that will likely not arrive at the same abstraction to services and integration between the IoT and cloud as well as interoperability in between additional local clouds, which may be needed.

3. Research Approach

The research approach was based on a case-study methodology proposed by Yin [16] with "a linear but iterative process" (p1) comprising: planning, design, preparation, data collection, data analysis and sharing of results. Alternative methodologies, such as action research or participatory design, would have also been possible to use, but might have required more time, effort and action from the researchers. The research question is "how much improvement has been made and how has the company managed to maintain a high development pace and efficiency throughout the development and simultaneous operation of the cloud service platform?", and the "how" indicates that a case-study methodology is appropriate to use. In addition, a case-study methodology was suitable as the researchers did not directly participate in the work that led to multiple contexts and customer cases.

The case-study company, Adage AB, is a high-tech micro SME located in northern Sweden and is very active on the northern European market. The company was founded in 2005 and has recently had an annual turnover of approximately 60 k€ and 110 k€ in 2016 and 2017, and expected in 2018 is 220 k€. Further, the company has one employee (the company currently uses development resources from its mother company BnearIT AB). The company provides services and products and participates in small and large research and development projects. Further, the company also provides cloud services and Industrial Product-Service Systems (IPS^2) [17].

The planning of the case study was crafted at the very start when the provider did the initial projects related to the construction and real estate contexts during 2010 and recycling management the year following. The focus on these three areas is due to that they all have similar technology problems and were developed on the same initial platform thinking (and later adapted to and re-built on the Arrowhead Framework) and that they were the starting points with interesting commercial outlooks. The planning was further updated during 2014 as a new business model was applied (IPS^2) to the recycling management context and the business volume increased rapidly. The intent was to follow the contexts on a regular basis and conclude the complete case study by 2017. The design of the case study included: formulating the study question, stating the study proposition: how has the company needed to change in terms of, e.g., the below, in order to improve and keep a high development pace and efficiency:

- Technology
- Development processes/practices
- Competencies and skills
- Organization
- Management
- Infrastructure
- Service and support

Most of these areas were initially predefined, except for management and infrastructure that were added later, based on practical experience from innovation and change management in companies. Further, the unit of analysis was the organizational level at the company. In addition, explanation building will be used to link the data to the propositions. It was decided that the criteria for interpreting the case study's findings would be made via rival explanations based on Patton's [18] approach, balanced defensively and offensively. The researchers wish to investigate how the company's set-up and organization has been affected when changing the development strategy and business model towards an IPS^2. The generalizability of the (multi-usable) cloud service platform is of great interest as well.

In order to prepare for the case study, a number of presentations, architectural/technical specifications, technical plans and marketing documents were analyzed. Further, a number of questions were presented to the company and ideas for changes and additional customer pain points were fed back to them as well. The initial data collection was done through semi-structured interviews [19,20] combined with workshops [21]. Further, intermediary data collection was conducted through multiple interviews and workshops, and the final data collection was accomplished via a workshop. Semi-structured interviews were used, with open-ended questions [19] allowing the respondents to give detailed answers and the possibility to add extra information where deemed necessary [20]. The duration of the interviews was between one and two hours, and the duration of the workshops was approximately two hours. In order to strengthen the validity of the study, the collected data were displayed using a projector during the interviews and workshops, allowing the respondents/participants to immediately read and accept the collected data. Subsequently, the collected data were displayed and analyzed using matrices (cf. [22]). The analyzed data were finally summarized into a matrix, and the findings categorized according to the areas of concern, i.e., development pace and efficiency gains, development process/practices, technology, competences and skills, organization, management, infrastructure, and service and support. Finally, the results were shared with the provider and some of their customers (as well as with a broader audience through this paper). During the case study the part of the research process (i.e., design/preparation/data collection) was iterated, since it was realized that more could be explored.

4. A Multi-Usable Cloud Service Platform and the Case Study's Three Contexts

This section will firstly outline the technical aspects of the cloud service platform architecture used, followed by the three contexts covered by the case study. The last context, i.e., recycling management, will be thoroughly described, as it is the most developed one.

4.1. Multi-Usable Cloud Service Platform—Architecture

The multi-usable cloud service platform from the company was initially developed using legacy technology and techniques. This was not deemed as efficient and scalable, and as the Arrowhead Framework and its components were made available during the Arrowhead project's progression (2014–2017)—the company re-engineered the complete solution and based it upon the Arrowhead Framework components as well as its proposed development processes and other related frameworks.

In brief, the Arrowhead Framework comprises the following main systems, sub-systems, procedures and methods [8,23]:

- **Mandatory core systems**—service registry, authorization, orchestration
- **Automation support systems**—plant description, configuration, system registry, device registry, event handler, quality of service manager, historian, gatekeeper, historian to historian secure data path, translation
- **Application systems**—application services
- **Deployment procedure**—secure bootstrapping of devices and systems into a local cloud, creation of Arrowhead Framework compliant systems, interfacing legacy systems, verification of compliance

Figure 1 outlines the new architecture, based on the Arrowhead Framework [23], for the company's cloud service platform. The architecture comprises the mandatory core systems and services, including an authorization system (for users and services, etc.), a service registry (where services can be registered and looked up), and an orchestration system (in order to maintain system connections), and an application systems and services part. The latter contains the actual business logic and provisioning, which differ for each context. The SOA principle is based on the foundation of the three 'L's, Lookup (discover/set presence), Loosely coupled (autonomy's and distributed components) and Late binding (dynamic system of system compositions) together with strictly defined service contracts and architectural methods.

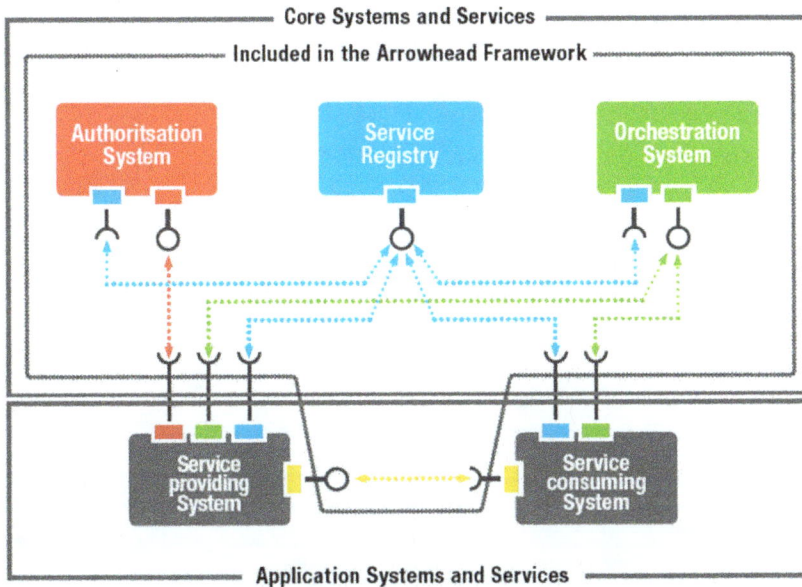

Figure 1. Outline of general architecture of Adage AB's multi-usable cloud service platform.

In further detail, the function calls and parameters used in Figure 1 are briefly outlined (for additional details see [8,23]). Input parameters are in normal text and output parameters in bold text:

- Authorisation (Service Discovery, **Authorisation Management, Authorisation Control**)—provides authorization and fine grained access rules to specific resources as well as configuration of access tickets in combination with use of authentication mechanisms such as, for instance, certificates and certificates handling
- Service Registry (**Service Discovery**)—provides service registry functionality based on DNS and DNS-SD standards. The architecture and solution handle self-registering services, which state their availability within the network. Consumers of specific information knows where they can find and use it. Producers notify their awareness. This handles the SOA-principles called lookup and late binding
- Orchestration (_ahfc-servprod types, Authorisation Control, Service Discovery, Orchestration Push, **Orchestration Store, Orchestration Capability, Orchestration Management**)—provides service consuming systems with consumption patterns as well as end-point information of the produced services that are to be consumed. The function provides the possibility to combine a system-of-system collaboration that fulfill the current need for a specific situation. Further, this handles the SOA-principles called late binding and loosely coupled.

4.2. Construction and Pre-Fabrication Context

Concrete pre-fabrication production is a very traditional craft which extensively builds on the hands-on experience of the producer. The same goes for builders that make concrete platforms for building houses or other structures. Common approaches for analogue sensing when the concrete pre-fabs or platforms are hardened and ready for use are to use the fingers to sense, using nails to estimate the hardness, or use other rules-of-thumb regarding what temperature or time has elapsed since the concrete moulds were filled, or which sand quality that was used in the concrete. This analogue setting is not optimal for development and also makes knowledge transfer time-consuming, e.g., when an adept is to take over from an experienced pre-fab maker or builder. In order to overcame this analogue setting and make it digital, the company has developed a sensor-bridge-cloud system with sensors embedded into the wet concrete. The sensor information is used as input to an algorithm which will let the pre-fab or builder management know the temperature and the humidity in the mould. The algorithm further calculates the hardening process of the concrete based on the time, humidity, temperature, type of concrete as well as additional parameters. With this information, the pre-fab maker or builder gets vital information if the ambient settings are optimal—and the algorithm can suggest optimizations by adjustments of time, temperature and humidity settings. This new capability both enables the pre-fab maker and builder to validate the quality of the set concrete and to know when to remove the mould(s). Thus, using collected data and the algorithm allows use of facts rather than estimations or rules-of-thumb to reduce/optimize the production time—which directly affects revenue.

4.3. Real Estate Context

Long-term owners of real estate, and in particular of larger commercial and multi-tenant apartment buildings, need to maintain the buildings in order to avoid deterioration (i.e., asset value is lost), uphold the customer value (be able to collect rent) as well as to keep the buildings healthy and comply to the societal requirements of energy efficiency and environmental consciousness/certifications. The company has for the real estate context developed an offer which measures, for instance, the moisture level in the exterior walls. This is needed for early detection of excessive moisture levels and thus a need to take action in order to avoid a complete renovation of the wall. A complete renovation of a wall far exceeds the cost of taking preventative measures at an early stage. The company's offer collects the data using sensors, and then the data is collected and analyzed in a central cloud service platform. This can be compared to commercial solutions with passive RFID tags that require a person to walk around and use a scanner. The benefit with this solution is that the batteries in the passive RFID tags will last for a long time. However, it is not a really scalable and cost-efficient solution if the real estate portfolio comprises several hundreds or thousands of buildings scattered over several cities or countries.

The company's offer can be extended with sensors; e.g., for water/leakage/flooding in basements or laundry rooms, status of doors (open/shut/number of passages), temperature (to detect open outer doors or fires), etc. Thus, actions that trigger the sensors can invoke immediate/planned inspections, alarms or event logging.

4.4. Recycling Management Context

The company provides a recycling IPS[2] to municipalities and companies who are responsible for managing recycling containers. The containers are mainly for households and are used to deposit glass, plastics, paper, packaging materials and metals. The main idea is to optimize the emptying of these containers as they are almost filled up (but not over-filled) to avoid people putting what is returned outside of the containers (i.e., littering) when they are over-filled. Further, the waste management company wishes to avoid emptying containers that do not need to be emptied. Thus, the waste management company wants to empty the containers on-demand and to minimize unnecessary transports with trucks in areas with high traffic loads as well as in rural areas (long distances). Figure 2 outlines a general overview of the IPS^2, of which the aim is to provide decision-making information to the customers with high availability.

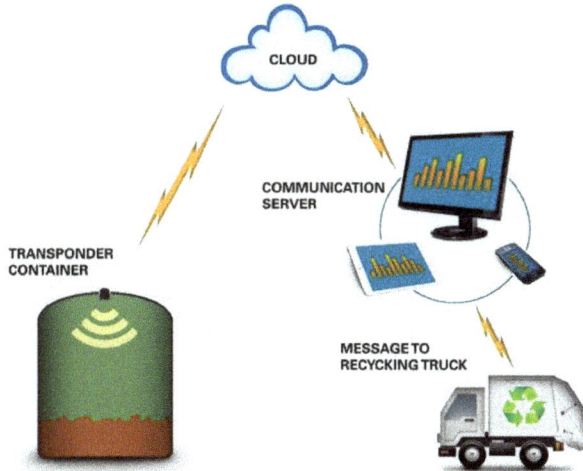

Figure 2. General outline of the Industrial Product-Service Systems (IPS^2) for optimizing recycling management.

Further, Figure 3 shows how the architecture is used/adapted for the recycling management context. On the service-providing system side, the containers and their sensors are connected to the cloud service platform and on the service-consuming system side, the output from the cloud service (i.e., decision-making information) can be consumed/accessed through the browser in a mobile phone, notepad or lap-top, etc. or be further processed in other information systems on the customer side.

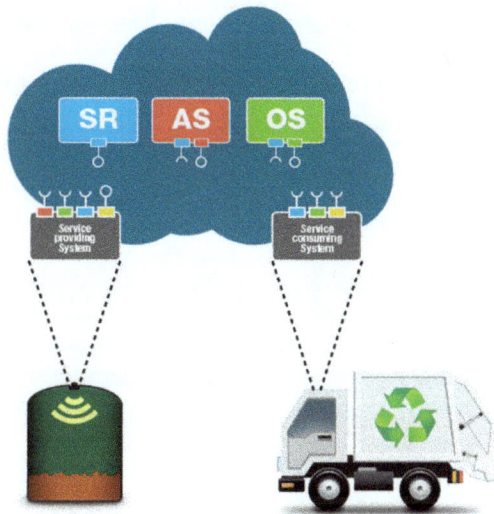

Figure 3. Architecture used/adapted for a recycling management context.

The adaption of the architecture for the construction and pre-fabrication as well as real estate contexts are made in the same manner, re-using/sharing all the business functionality from the recycling management context and its implementation.

The IPS^2 is built on the industrial internet and IoT paradigms. The IPS^2 uses sensors to measure the filling level within containers, vibrations (to indicate emptying), as well as temperature (to adjust filling-level measurements and indicate/detect possible fires) and other wanted parameters. Further, the IPS^2 pre-processes data to minimize data flows, uses wireless communications (mainly mobile networks or WLANs), provides a cloud service for data collection and analysis of big data, and performs additional wanted data mining and visualization of results (such as which containers should be emptied). The IPS^2-components attached to the grid are powered by green electricity (hydropower) and the cloud service is run in green data centers (powered by green electricity and cooled by the low-temperature air in northern Sweden).

Figure 4 shows how some of the decision-making information is visualized for managers at the customers. Consequently, the customers need to change their processes and management to be able to fully benefit from the decision-support provided—in order to achieve good optimization.

Using the information provided in Figure 4, the managers concerned with managing the recycling get information about which containers need to be emptied (or will soon need to be emptied). Further, the managers are supported when planning optimal routes and dispatching an optimal number of trucks. Figure 5 provides graphs about the filling details when the managers click on the containers in Figure 4.

In addition, to make it simple, another status indicator is possible to add in the decision-making information as well. Figure 6 is commonly configured so that the green container is less than 50% filled up, the yellow between 50–80% and the red at +80% (thus indicating that emptying is needed soon). A prognosis for which containers will reach +80% within 4–5 days can be calculated based on the filling patterns.

Further, a graphic or numeric filling report can be issued or retrieved for each customer's containers and filling-levels. In addition, commonly the filling-levels green/yellow/red can be customized at wanted thresholds for each customer.

Customers get improved (realtime) information with high availability regarding the filling-level of containers, and are thus able to better plan emptying schedules and dispatch an optimal number of trucks (and avoid over-filled containers). The latter is important, as the customers are responsible for cleaning up any litter or recyclables that are left outside of the over-filled containers. This takes time, and sometimes a lot of time, when the emptying trucks should instead be on the move, performing their actual task. Realtime monitoring of the filling-levels provides input for planning emptying, and using historic data and analytics a prognosis can also be made for when the containers will reach +80% filling-level. According to one major customer concerned with the glass containers, they have made large improvements compared to in the past when containers where emptied at an average 55% filling-level and every 12th container was over-filled. Thus, the customers can use an optimal number of trucks (which has a large impact on profitability), take other suitable contracts that can be combined when they are better able to plan the recycling business, drive less in areas with high traffic loads, and drive less in rural areas with large distances in between containers.

Thus, the total outcome, enabled by the IPS^2, supports the circular economy in an environmentally friendly manner.

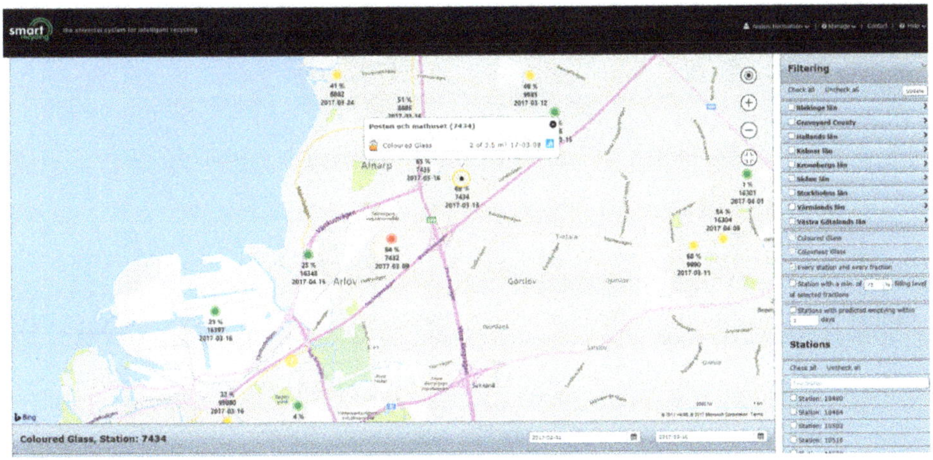

Figure 4. Visualization of a management view—high-level decision-making.

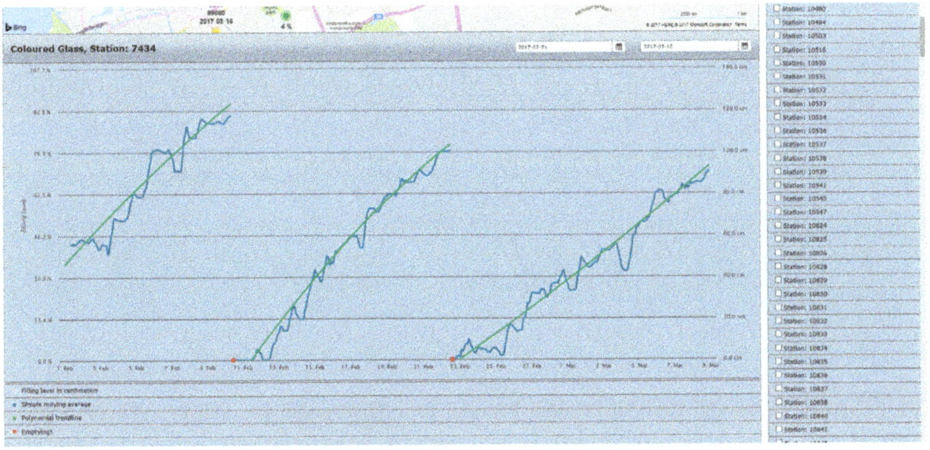

Figure 5. Decision-making graphs—drill-down on container filling levels.

Figure 6. Simple decision-making—visualization of container filling-levels (green/yellow/red).

5. Results

Regarding the degree of improvement and how the company has managed to keep a high development pace and efficiency throughout the development and simultaneous operation of the cloud service platform, the results are listed in Table 1. The findings are the outcome of the workshops and interviews conducted during the case study. The findings are attributed to: development pace and efficiency gains (PEG), development process/practices (DPP), technology (T), competencies and skills (CS), organization (O), management (M), infrastructure (I), and service and support (S).

Table 1. Improvement and how a high development pace and efficiency has been accomplished at Adage AB.

#	Area	How Much and How
1	PEG	From start when using traditional/legacy development processes/methods/tools, etc. until later stages—a 50–75% improvement regarding pace and efficiency. Thus, a significant saving of effort, time and costs.
2	DPP	Usage of the Arrowhead Framework [7,21] and its requirement engineering and development process, as well as usage of strict Service-Oriented Architecture (SOA) principles, has enabled streamlined development and cooperation.
3	T	Usage of the Arrowhead Framework to abstract IoT devices and sensors as services saves a lot of time and effort. Further, re-use and refinement of the data collection and analytic platform, together with re-use of analytic tools and frameworks, has further hastened development and improved efficiency. The addition of secure remote configuration and updating of the cloud software and sensor software has been key to scalable technical management and operational efficiency.
4	CS	Firstly, whole new businesses (and customer segments) unfamiliar to us had to be learnt. We were also used to mainly providing products and services, and had to learn more about complex business models. Secondly, we needed to learn more about: cloud services, SOA, Arrowhead Framework, sensors, robust data and mobile communications, industrial internet/IoT, big data, data modeling and data analytics, and IT/information/cyber-security. In particular, the business parts, data modeling, data analytics and security have been challenging. Thus, this has been very challenging—but on the other hand developed the company considerably.
5	O	It was necessary to change to a new business model i.e., IPS^2, from a product with a services model. When changing the business model, with subscriptions based on 36–60 month contracts and the pricing depending on the volume, the business took off and the volumes rapidly started to grow. Further, we have started to set up partner networks in order to scale up the business and sales/marketing efforts. In addition, we had to win the trust of the customers by firstly knowing their business, and then having an offer that they found attractive.
6	M	Internally, we have had to add missing competencies to the board of directors, and use long-term financial planning and strict cash-flow to not impede the fast pace of customer installations.
7	I	We moved from an own operated cloud service to a green-powered cloud platform service provider, which has allowed us to focus more on technical- and business development instead of operations. Further, this is much more stable and we do not have to have the same number of people concerned with operations.
8	S	We have learnt a lot by doing service and support, set up a service desk and support organization, contracted service partners, and have consequently developed e.g., health indices for sensors (i.e., battery level and function) and other equipment in order to reveal maintenance need for physical installation and ensure high availability of the decision-making information to the customers.

The company estimates that it has saved between 50–75% development time, for each of the three contexts, by using the Arrowhead Framework combined with strict SOA-principles instead of using "more traditional/legacy" development processes, methods, practices, and tools, etc. By following the principles and processes in the Arrowhead Framework, it has been easy and straightforward for requirement engineering and development activities and further resulted in fewer misunderstandings. Further, there is improvement for each iteration as additional experience and re-use of components, analytical tools and frameworks etc., provide effect over time. The savings estimation is based on comparisons of development plans and efforts spent. The savings estimations for 6 larger projects are: 50%, 55%, 60%, 60%, 70 and 75%—i.e., in between 50–75%. To corroborate this saving, previous research conducted by Delsing et al. [15] shows that the use of the Arrowhead framework saves development time/efforts, related to industrial IoT and cloud services, in the range of approximately 70–80%. Thus, this ought to be of interest also for other companies and organizations involved in development activities related to the industrial internet and IoT.

The eight areas in Table 1 are all interconnected as they rely on, support or depend on each other. The areas #1, 2, 3, 4 and 6 are directly affecting the development pace and efficiency whereas areas #7 and 8 indirectly contribute. Finally, area #5 is driving the rest and at the same time supported by the rest.

6. Analysis

The pace and efficiency regarding the development of the multi-usable cloud service platform have been significantly improved compared to previous reliance on more traditional development methods and tools. In this regard, the Arrowhead framework has been the key enabler to the improvement—although a number of the areas in Table 1, increased experience and re-use of tools etc. have also contributed with approximately 20–25% of the **50–75% improvement** at the company. However, re-use of analytical tools and frameworks would, of course, have been possible anyway. In order to be successful with the customer offers, the company has experienced that the development of technology, business model, business set up and organization needed to be synchronized and go hand in hand. The development-, technical- and infrastructure-related matters were perceived as easier to overcome compared to the business model-, business set up, organizational (including acquisition of skills and competencies), management, and service and support-related ones. One explanation for this is that the company was a high-tech micro SME with mainly engineers employed initially—and thus more interested in the engineering-related areas. This can be an important lesson for others considering taking on a similar challenge.

To conclude the analysis, Chesbrough [24] posits that the choice of business model is key, as "a mediocre technology pursued within a great business model may be more valuable than a great technology exploited via a mediocre business model" (p. 355).

7. Discussion and Conclusions

The paper has contributed to literature with a case of a multi-usable cloud service platform, developed and used by a micro SME, outlining that the micro SME has improved its development pace and efficiency by 50–75%. This paper also describes how this process has been accomplished. Of particular interest is to notice the main changes i.e., use of the Arrowhead Framework as well as strict use of SOA-principles. Further, the areas in Table 1 have contributed to increased development experience and re-use as well as improved knowledge of customers, customer processes, and the overall system. The possibility to abstract IoT devices from various vendors as services provides a great advantage, both during the development/integration phase and later during the operations/maintenance phase.

Further, the paper has given input to practice by indicating that a number of changes are needed to improve the development pace and efficiency. In particular, re-use and refinement of the data collection and data analytic platform plus re-use of analytic tools and frameworks rendered a good result. However, this would have been possible in many other contexts as well without using the Arrowhead Framework. Regarding scalable technical management, remote configuration and updating of cloud service platform and sensor software are key for operational excellence.

In addition, the paper has provided lessons for management, as the micro SME also changed its business model to an IPS^2 from a product with services. The business model change, together with other necessary organizational-, managerial- and infrastructural changes, combined with adding missing competencies, invoking long-term financial management and strict cash-flow, setting up a service and support organization, was demanding and required a lot more effort and resources than initially anticipated. Thus, to be successful in the innovations, the technical aspects need preferably to go hand in hand with business and organizational aspects. This will create stronger customer value and, consequently, an attractive customer offer.

Regarding the case study's proposition, it is clear that the company has had to change in all aspects listed (and probably more that were not discovered during the case study). Making such a number of changes during a few years requires management support, funding (the company has received considerable additional funding from its owners during the duration of the case study), a technical vision, inclination for business development, and trying new business models to learn new markets and embrace change management. Making a lot of changes is hard, and requires a lot from an organization. However, compared to larger organizations, it is likely easier for a micro SME to make a lot of changes.

Concerning the rival explanations, in terms of technology and organization, the company looks completely different after the case study. The new business model used (IPS^2) has impacted the organizational setup and need to change the underlying technology to an efficient multi-usable cloud service platform. As a rival explanation—would it have been wise and better to keep using traditional/legacy development processes, methods, practices, tools, etc. compared to moving into the Arrowhead Framework and SOA-principles? For the case study in question—the answer is no.

The research question "how much improvement has been made and how has Adage AB managed to maintain a high development pace and efficiency throughout the development and simultaneous operation of the cloud service platform?" has been answered above. An improvement of 50–75% should be of great interest for other companies and organizations to investigate as well. Further, regarding generalizability, the company's use of the Arrowhead Framework is possible for others as well, since most of it is available as open-source [23]—and possible to download, test, and use. In addition, the managerial lessons i.e., that a lot of changes need to go hand in hand to successfully get new offers to the market and stay competitive, can also be of a general interest for other organizations planning for cloud service platforms.

In terms of sustainability, the paper indicates that considerable improved economic sustainability can be reached for the provider through higher efficiency in engineering and development efforts. Further, as the cloud service platform is used for instance in optimization of recycling management as well as monitoring in real estate, the cloud service platform also has an impact on environmental and social sustainability.

The company will, according to its management, continue to develop the multi-usable cloud service platform and the whole setup, and look to continue to develop their offers and business modeling. Another demanding business model of interest for the future is to provide a function or Functional Product [25–27].

Acknowledgments: The research has partly been funded by the VINNOVA VinnVäxt ProcessIT Innovations R&D Centre at Luleå University of Technology, Sweden. Further, the authors would like to thank all respondents for their valuable input and time.

Author Contributions: John Lindström has been the main author supported by the three co-authors Petter Kyösti, Anders Hermanson and Fredrik Blomstedt. Anders Hermanson and Fredrik Blomstedt have provided data and access to the case study descriptions as well as their figures.

Conflicts of Interest: The authors declare no conflicts of interest.

References

1. Demirkan, H.; Delen, D. Leveraging the capabilities of service-oriented decision support systems: Putting analytics and big data in cloud. *Decis. Support Syst.* **2013**, *55*, 412–421. [CrossRef]
2. Hashem, I.A.T.; Yaqoob, I.; Anuar, N.B.; Mokhtar, S.; Gani, A.; Khan, S.U. The rise of "big data" on cloud computing: Review and open research issues. *Inf. Syst.* **2015**, *47*, 98–115. [CrossRef]
3. Derhamy, H.; Eliasson, J.; Delsing, J.; Priller, P. A survey of commercial frameworks for the Internet of Things. In Proceedings of the IEEE 20th Conference on Emerging Technologies & Factory Automation (ETFA), Luxembourg, 8–11 September 2015; pp. 1–8.
4. Amazon Web Services. Available online: https://aws.amazon.com/ (accessed on 23 March 2017).
5. Microsoft Azure. Available online: http://azure.mocrosoft.com/ (accessed on 27 March 2017).

6.   Google Cloud Platform. Available online: https://cloud.google.com/ (accessed on 23 March 2017).
7.   IBM Bluemix. Available online: https://www.ibm.com/cloud-computing/ (accessed on 23 March 2017).
8.   Delsing, J. (Ed.) *IoT Automation—Arrowhead Framework*; CRC Press: Boca Raton, FL, USA, 2017.
9.   Dyche, J. Data-as-a-Service, Explained and Defined. SearchDataManagement.com. Available online: http://searchdatamanagement.techtarget.com/answer/Data-as-aservice-explained-and-defined (accessed on 22 March 2017).
10. Birk, D.; Wegener, C. Technical issues of forensic investigations in cloud computing environments. In Proceedings of the IEEE Sixth International Workshop on Systematic Approaches to Digital Forensic Engineering (SADFE), Oakland, CA, USA, 26 May 2011; pp. 1–10.
11. Jensen, M.; Schwenk, J.; Gruschka, N.; Iacono, L.L. On technical security issues in cloud computing. In Proceedings of the IEEE International Conference on In Cloud Computing CLOUD'09, Bangalore, India, 21–25 September 2009; pp. 109–116.
12. Li, M.; Sun, X.; Wang, H.; Zhang, Y.; Zhang, J. Privacy-aware access control with trust management in web service. *World Wide Web* **2011**, *14*, 407–430. [CrossRef]
13. Sun, X.; Wang, H.; Li, J.; Zhang, Y. Injecting purpose and trust into data anonymization. *Comput. Secur.* **2011**, *30*, 332–345. [CrossRef]
14. Fylaktopoulos, G.; Skolarikis, M.; Padadopoulos, I.; Goumas, G.; Sotiropoulos, A.; Maglogiannis, I. A distributed modular platform for the development of cloud based applications. *Future Gener. Comput. Syst.* **2018**, *78*, 127–141. [CrossRef]
15. Delsing, J.; Eliasson, J.; de Venter, J.; Derhamy, H.; Varga, P. Enabling IoT automation using local clouds. In Proceedings of the IEEE World Forum on Internet of Things, Reston, VA, USA, 12–14 December 2016; pp. 501–507.
16. Yin, R.K. *Case Study Research: Design and Methods*; Sage Publications: Thousand Oaks, CA, USA, 2003.
17. Meier, H.; Roy, R.; Seliger, G. Industrial Product-Service Systems—IPS². *CIRP Ann. Manuf. Technol.* **2008**, *59*, 607–627. [CrossRef]
18. Patton, M.Q. *Qualitative Research and Evaluation Methods*, 3rd ed.; Sage Publications: Thousand Oaks, CA, USA, 2002.
19. Kvale, S.; Brinkmann, S. *InterViews: Learning the Craft of Qualitative Research Interviewing*; Sage Publications: Thousand Oaks, CA, USA, 2009.
20. Fontana, A.; Frey, J. Interviewing. In *Handbook of Qualitative Research*; Denzin, N., Lincoln, Y., Eds.; Sage Publications: Thousand Oaks, CA, USA, 1994.
21. Remenyi, D. *Field Methods for Academic Research: Interviews, Focus Groups & Questionnaires in Business and Management Studies*, 3rd ed.; Academic Conferences and Publishing International Limited: Reading, UK, 2013.
22. Miles, M.; Huberman, M. *An Expanded Sourcebook—Qualitative Data Analysis*, 2nd ed.; Sage Publications: Thousand Oaks, CA, USA, 1994.
23. Arrowhead Framework. Available online: http://www.arrowhead.eu/arrowhead-wiki/ (accessed on 23 March 2017).
24. Chesbrough, H. Business Model Innovation: Opportunities and Barriers. *Long Range Plan.* **2010**, *43*, 354–363. [CrossRef]
25. Alonso-Rasgado, T.; Thompson, G.; Elfstrom, B.-O. The design of functional (total care) products. *J. Eng. Des.* **2004**, *15*, 515–540. [CrossRef]
26. Lindström, J.; Plankina, D.; Nilsson, K.; Parida, V.; Ylinenpää, H.; Karlsson, L. Functional products: Business model elements. In Proceedings of the Product-Service Integration for Sustainable Solutions: Proceedings of the 5th CIRP International Conference on Industrial Product-Service Systems, Bochum, Germany, 14–15 March 2013; Horst, M., Ed.; Springer Science + Business Media B.V.: Berlin/Heidelberg, Germany, 2013; pp. 251–262.
27. Lindström, J.; Karlberg, M. Outlining an overall Functional Product lifecycle: Combining and coordinating its economic and technical perspectives. *CIRP J. Manuf. Sci. Technol.* **2017**, *17*, 1–9. [CrossRef]

applied
sciences

MDPI

Article

An Integrated Open Approach to Capturing Systematic Knowledge for Manufacturing Process Innovation Based on Collective Intelligence

Gangfeng Wang [1,*], Yongbiao Hu [1], Xitian Tian [2], Junhao Geng [2], Gailing Hu [3] and Min Zhang [2]

[1]   Key Laboratory of Road Construction Technology and Equipment of MOE, School of Construction Machinery, Chang'an University, Xi'an 710064, China; hybiao@chd.edu.cn
[2]   School of Mechanical Engineering, Northwestern Polytechnical University, Xi'an 710072, China; tianxt@nwpu.edu.cn (X.T.); gengjunhao@nwpu.edu.cn (J.G.); zhangmin0907@mail.nwpu.edu.cn (M.Z.)
[3]   School of Mechanical Engineering, Xi'an Jiaotong University, Xi'an 710049, China; hugl@mail.xjtu.edu.cn
*   Correspondence: wanggf@chd.edu.cn or gangfengwang@outlook.com; Tel.: +86-29-8233-4586

Received: 3 February 2018; Accepted: 22 February 2018; Published: 27 February 2018

Abstract: Process innovation plays a vital role in the manufacture realization of increasingly complex new products, especially in the context of sustainable development and cleaner production. Knowledge-based innovation design can inspire designers' creative thinking; however, the existing scattered knowledge has not yet been properly captured and organized according to Computer-Aided Process Innovation (CAPI). Therefore, this paper proposes an integrated approach to tackle this non-trivial issue. By analyzing the design process of CAPI and technical features of open innovation, a novel holistic paradigm of process innovation knowledge capture based on collective intelligence (PIKC-CI) is constructed from the perspective of the knowledge life cycle. Then, a multi-source innovation knowledge fusion algorithm based on semantic elements reconfiguration is applied to form new public knowledge. To ensure the credibility and orderliness of innovation knowledge refinement, a collaborative editing strategy based on knowledge lock and knowledge–social trust degree is explored. Finally, a knowledge management system *MPI-OKCS* integrating the proposed techniques is implemented into the pre-built CAPI general platform, and a welding process innovation example is provided to illustrate the feasibility of the proposed approach. It is expected that our work would lay the foundation for the future knowledge-inspired CAPI and smart process planning.

Keywords: manufacturing process innovation; computer-aided innovation; open innovation; collective intelligence; knowledge management; knowledge-based engineering

1. Introduction

In today's rapidly changing market landscape, regardless of any product industry, technological innovation has been regarded as an important factor for manufacturing enterprises to ensure future competitive advantage. As a basic form of technological innovation, manufacturing process innovation is the key guarantee for the R & D final realization of new products [1–3], especially in the field of complex equipment, such as aircraft, aerospace, automobile, construction machinery, and so on [4–7]. Because the structure of the world economy has undergone significant changes, with demand for energy saving and environmental protection becoming increasingly urgent [8–10], developing countries need to transform and upgrade their manufacturing industries with process innovation to reduce energy consumption and achieve sustainable development; developed countries, accordingly, are trying to guide and accelerate the global return of manufacturing industries by means of process innovation [11,12].

However, manufacturing enterprises have long encountered a variety of problems in the implementation of process innovation. These problems are mainly manifested in the difficulty of innovation, the poor effect, and the low success rate [13,14]. Generally speaking, the new manufacturing process technologies—especially sustainable process technologies—often entail long-term, complex, experimental, and higher-risk development efforts [15–17]. Industrial innovation survey data shows that the lack of technical staff and relevant innovation knowledge is one of the prime reasons for the termination or failure of innovation activities [3,13,18]. In fact, manufacturing process innovation is a cross-industry and interdisciplinary type of complex system engineering, which requires not only domain experts with multidisciplinary knowledge, but also technical or management personnel of manufacturing sites with process know-how [14,19]. Nevertheless, the empirical knowledge existing in these scattered owners has not yet been effectively organized according to innovation design procedure and cannot currently be applied to Computer-Aided Process Innovation (CAPI) [3,20].

It is recognized that reasonable and efficient innovation knowledge capture is the foundation for the effective innovation knowledge application, and it is regarded as one of the core requirements for smart innovation engineering of the Future Industry 4.0 [21–24]. Although several pre-research works exist in process innovation knowledge management and CAPI framework [2,3,25], there is still a lack of an integrated approach to effectively capturing systematic process innovation knowledge under the open innovation paradigm. The open process innovation knowledge capture is, essentially, a process of effective combination of knowledge owners' collective intelligence [20,26]. It will be able to match the characteristics of process innovation knowledge and make full use of the wisdom of multidisciplinary and multi-sectoral personnel, so as to meet the needs of CAPI-oriented knowledge organization.

Consequently, our goal in this research work is to construct an open knowledge capture approach, which can obtain structured, formalized, and systematic innovation knowledge from open environments and thus support manufacturing process problem-solving. By building an open knowledge–social community and considering multi-type knowledge organization and evolution in the process of knowledge-inspired innovation design, a novel holistic paradigm of process innovation knowledge capture based on collective intelligence (PIKC-CI) and the corresponding knowledge processing approach are explored. Accordingly, an open knowledge capture system for manufacturing process innovation (*MPI-OKCS*) is constructed in this paper, in order to implement the proposed method for practical application.

The remainder of this paper is organized as follows. In Section 2, some related works about innovation-oriented knowledge capture and CAPI are reviewed. Section 3 presents the overall paradigm of PIKC-CI. Section 4 shows the detailed procedure of the proposed PIKC-CI method, mainly including multi-source knowledge fusion and collaborative knowledge refinement. Then, a prototype system *MPI-OKCS* is implemented in Section 5 and further studied, with a case application of welding process innovation knowledge capture by using the mentioned method. The last section concludes this paper with some implications for future research.

2. Literature Review

2.1. Innovation-Oriented Knowledge Capture and Management

As is commonly recognized, knowledge is an essential asset for organizations and plays a crucial role in innovation; from another perspective, innovation can be regarded as the knowledge-based creation and the knowledge-based outcome [27,28]. To focus this study, related research has been conducted in previous contributions to innovation knowledge management and knowledge-based innovative design. Esterhuizen et al. [29] explored how knowledge conversion can grow innovation capability maturity, and provided a framework for the use of knowledge creation processes as a vehicle to improve innovation. By exploring the complex relationships between knowledge management and innovation, Xu et al. [30] proposed an integrated approach to knowledge management for innovation, and developed a corresponding distributed prototype system. Bosch-Mauchand et al. [31] presented a novel approach to support the assessment of manufacturing process performance based on knowledge management integration. To effectively support systematic manufacturing process innovation, Wang et al. [32] presented an approach to principle innovation knowledge extraction from process patents.

In the knowledge-based economy, it is difficult for a single person or enterprise to have all the knowledge needed to achieve innovation. In the engineering field, open innovation is defined as the use of purposeful knowledge transfer in order to accelerate internal innovation and expand the application markets of external innovation [33,34]. Open innovation has recently become a new model of technological innovation because of its ability to combine internal and external collective intelligence [35,36]. Besides, the latest Web 2.0 technologies lay more emphasis on online collaboration and information sharing between users, and provide a technical basis for open knowledge capture and management. By combining open innovation strategy and Web 2.0 technologies, Hüsig and Kohn [37] introduced a new form of Computer-Aided Innovation (CAI)—"Open CAI 2.0".

2.2. Computer-Aided Process Innovation

Firstly proposed by J.A. Schumpeter from the perspective of economic development [38], process innovation received attention from both academic research and industry [19,20,39]. He believed that process innovation and product innovation constitute the technological innovation system of enterprises. The technological developments of information and communication technology (ICT) and innovation theory have provided a more structured knowledge-driven environment for technicians and market decision-makers [40–42]. Computer-based applications, such as CAD/CAE/CAPP/CAM, help users to achieve better solutions and hence to introduce better products, processes, and services to the diversified markets [17,43,44]. Meanwhile, the combination of innovation theory and ICT to support technological innovation has become a new research category known as CAI [40]. However, from a practical point of view, most of the current methods or tools of innovative design are more suited for product innovation than process innovation; sometimes they not only do not enhance the process innovation ability of manufacturing enterprises, but also even have some negative effects on production efficiency [39,45]. It is necessary for us to realize that process innovation and product innovation are quite different. In general, the process of process innovation covers a wider technical field, involves more participants, and suffers more realistic constraints. Actually, the traditional computer aided tools of the manufacturing process (e.g., CAPP/CAM) mainly focus on improving the efficiency and standardization of process design and management [23,43,46], rather than creating or improving process methods, and therefore cannot systematically enhance the development level of the manufacturing process in enterprises [2,3,15].

In recent years, some domain research endeavors have been carried out into specific types of manufacturing process innovation by using the Theory of Inventive Problem Solving (TRIZ) [42] and knowledge-based engineering [23]. Cakir and Cilsal [47] introduced a TRIZ-alike matrix-based access system and established a knowledge database for various contradictions of chip removal process. Duflou and D'hondt [48] applied TRIZ principles of physical conflict, resolving to improve the performance of single point incremental forming. By focusing on the semiconductor industry, Sheu et al. [49] developed a suitable contradiction matrix and corresponding inventive principles for that particular industry based on chemical–mechanical processing patents. With the development of CAI and the requirements of manufacturing process problem-solving, the basic concept and framework for CAPI were presented by Geng, Tian, and Wang [2,3,25,50], with some specific application cases being used to illustrate the feasibility of structured/systematic process innovation design [20,32,51].

2.3. Summary

In summary, much research has been done regarding aspects of innovation design theory and methods, and innovation knowledge modeling and management; however, very little work has addressed systematic knowledge-driven process innovation design and CAPI. It's gratifying that the existing research results have shown the feasibility of structured process innovation with the computer-aided method.

Currently, CAI is developing towards a knowledge-driven, open, and systematic direction. As a branch of CAI, CAPI is more focused on solving manufacturing process problems, improving process methodologies, fostering whole process innovation design cycles, and even enhancing the overall manufacturing innovation capability of enterprises. Manufacturing process innovation knowledge, which exists in the entire lifecycle of process innovation, is used to support the correct implementation of process innovation activities, and to produce new process knowledge [2]. Obviously, the formalized knowledge capture and management is crucial to systematic CAPI, especially under the open innovation paradigm. Thus, this paper will mainly explore CAPI-oriented open innovation knowledge capture based on collective intelligence.

3. An Overall Paradigm for Innovation Knowledge Capture Based on Collective Intelligence

From the systems thinking perspective, the innovation realization of CAPI is essentially the process of capturing and applying process innovation knowledge to solve specific process problems with the support of innovation theories, methods, and tools. Problem solving is a complex intellectual activity based on high-order cognition, and innovative problem solving is considered to be the process of overcoming at least one obstacle that impedes the achievement of the desired goal [52]. Thus the problem-solving of process innovation actually mainly includes analysis and formulation of process problems, process conflict extraction and resolution, detailed design of process innovation schemes, and evaluation and optimization of the scheme. The innovation design procedure can be basically divided into four stages, as illustrated in Figure 1, and each stage needs the support from the corresponding type of innovation knowledge. According to the role of knowledge in manufacturing process innovation design, we divide innovation knowledge into several types, such as Process Contradiction Matrix, Manufacturing Scientific Effect, Innovative Scheme Instance, and so on [25]. The above types of knowledge are required to be explicit, structured and formalized descriptions, so as to stimulate the creative thinking of the process designers and facilitate the implementation of knowledge-inspired innovative design in the computer support environment. Although the designers and experts in the manufacturing field have strong process problem-solving experience and rich manufacturing knowledge, this discrete and unstructured knowledge cannot be directly and efficiently applied to innovation design, nor is it conducive to knowledge capture and accumulation in manufacturing enterprises. Thereby, we need to explore an approach that can contact appropriate knowledge holders and make full use of their collective intelligence to participate in knowledge capture activities.

From the practical point of view of collective intelligence, the effect of knowledge capture and accumulation based on community is better than that based on the company's organization structure, because it can better share and focus the knowledge topics; knowledge refinement based on peer collaboration is better than that based on expert-centered editing, because it can narrow the distance between knowledge [26,53]. Thus, a novel manufacturing process innovation knowledge capture paradigm based on collective intelligence is proposed, just as shown in Figure 1. In an open knowledge–social community, personal knowledge can be gradually transformed into public innovation knowledge through knowledge–social activities among participants. The procedure of knowledge capture basically includes three main steps: knowledge contribution (KC), knowledge fusion (KF), and knowledge refinement (KR). Firstly, knowledge topics can be published according to the requirements of current manufacturing process innovation. Then interested users are gathered into a group through knowledge–social relationships. In the knowledge–social community, they discuss the topics and manifest their knowledge using knowledge templates from the viewpoint of individual specialty and experience. Then, the knowledge capture system will integrate this personal knowledge into the public knowledge fusion units under the semantic constraints of domain ontology. Thus, the knowledge fusion units will be iteratively edited and refined into formalized and systematic knowledge by refinement group. Subsequently, the captured process innovation knowledge can be effectively applied in the stage of innovation design.

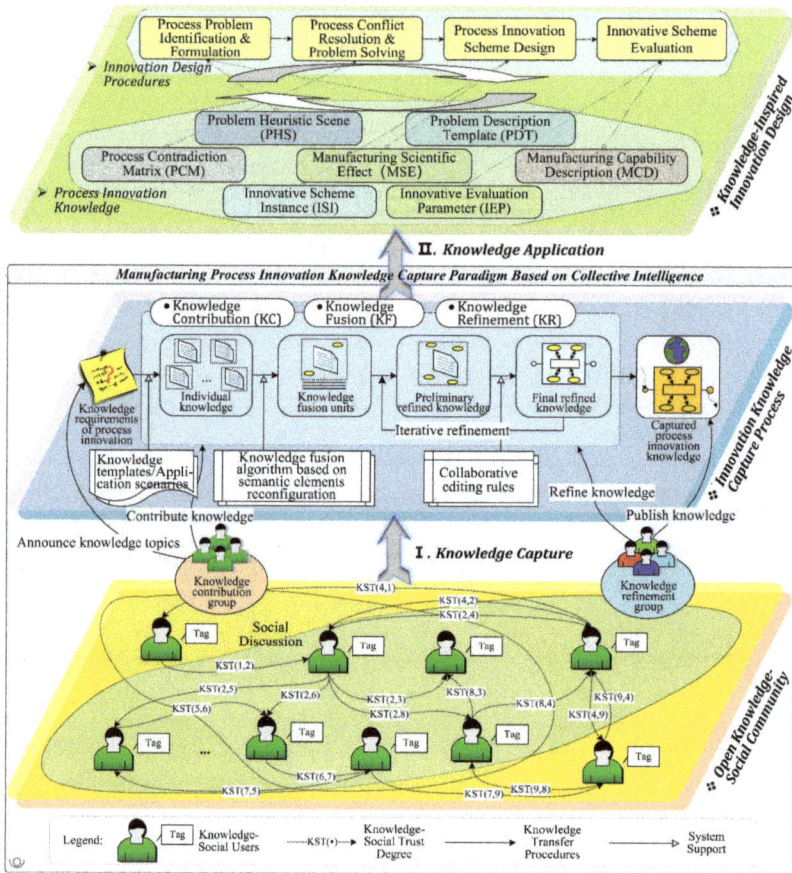

Figure 1. A novel process innovation knowledge capture paradigm based on collective intelligence.

4. The Proposed PIKC-CI Method

As revealed in Figure 1, we know that several knowledge activities, KC, KF and KR, are all needed for the integrated PIKC-CI method. Among them, multi-source knowledge fusion and collaborative knowledge refinement are the crux of the efficient innovation knowledge capture. In this section, the detailed approaches for multi-source innovation knowledge fusion, based on semantic elements reconfiguration, and collaborative innovation knowledge refinement, based on knowledge–social trust degree, are successively explored from the perspective of knowledge processing and transfer.

4.1. Multi-Source Innovation Knowledge Fusion Based on Semantic Elements Reconfiguration

For the convenience of detailed elaboration, this sub-section first presents the relevant definitions for process innovation knowledge and its fusion process.

Definition 1. *Manufacturing process innovation-oriented knowledge network is a set of spatial knowledge structure, formally represented as*

$$PIK_\Omega = \{KN, CTR, U\} \tag{1}$$

where KN is a set of multi-type process innovation knowledge units, CTR is a set of knowledge contextual relevance for specific process innovation scenarios, and U is a set of social-wiki users involved in knowledge capture.

The hierarchical structure of the process innovation-oriented knowledge network is shown in Figure 2 and formally defined as follows.

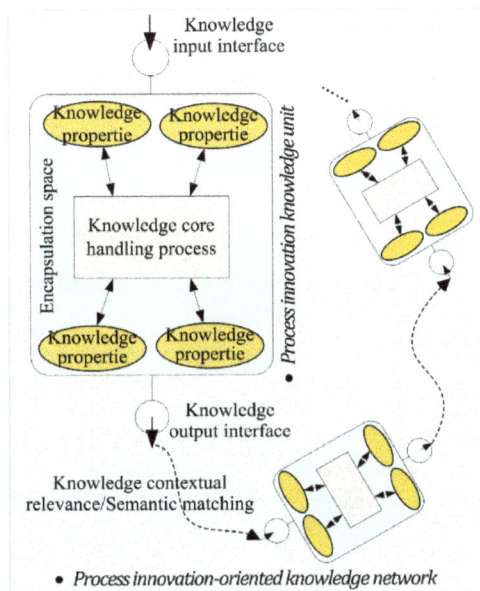

Figure 2. Schematic diagram of process innovation-oriented knowledge network.

Definition 2. *Process innovation knowledge unit is a local capability unit that has the ability to solve certain types of process problems and deliver information. It is defined as*

$$KN =< P, I_I, I_O, E, U > \tag{2}$$

where P is a set of knowledge properties, I_I and I_O represent the sets of knowledge input interface and knowledge output interface, respectively. E stands for an encapsulation space for complete knowledge units. Several types of innovation knowledge, $\Pi_{KN} = \{PHS, PDT, PCM, MSE, ISI, IEP, MCD\}$, are basically used in the innovation design process. Among them, PHS is the Problem Heuristic Scene, PDT is the Problem Description Template, PCM is the Process Contradiction Matrix, MSE is the Manufacturing Scientific Effect, ISI is the Innovative Scheme Instance, IEP is the Innovative Evaluation Parameter, and MCD is the Manufacturing Capability Description.

Definition 3. *Knowledge contextual relevance of manufacturing process innovation is further denoted by*

$$CTR = \{\langle kn, k, r, k', u \rangle | kn \in KN, k, r, k' \in \mathbb{O}, u \in U\} \tag{3}$$

where k, r, k' are ontological entities defined in process innovation domain ontology $\mathbb{O}$, *and r is a contextual relationship between k and k'.*

Definition 4. *Domain ontology* $\mathbb{O}$ *consists of a series of concepts and relationships that represent domain knowledge models. It is defined as*

$$\mathbb{O} := (C, R, \mathcal{E}_R, I_C) \tag{4}$$

where C and R are a set of classes and a set of relations, respectively; $\mathcal{E}_R \subseteq C \times C$ *represents a set of relationships between classes, which can be denoted as a set of triples* $\{\langle c, r, c' \rangle | c, c' \in C, r \in R \}$; *and* I_C *is a power set of instance sets of a class* $c \in C$.

The knowledge elements of process innovation knowledge are generally expressed in terms of domain terms or natural language descriptions. For example, process conflict parameters can be expressed as process parameters and their deformation, while process innovation principles can be expressed in natural language form. A knowledge element of natural language descriptions is composed of one or more propositions; a proposition is a complete semantic unit that contains terminology and predicate terms.

Definition 5. *Process innovation knowledge element represents a complete and indivisible knowledge unit in knowledge space* PIK_{Ω}. *It is defined as*

$$Ke = \{\Sigma, \Lambda, \Theta\} \tag{5}$$

where $\Sigma = \{t_1, t_2, \dots t_i, \dots t_n\}, t_i (i = 1, 2, \dots, n)$ *is terminology, and* $\Lambda = \{p_1, p_2, \dots p_j, \dots p_m\}, p_j (j = 1, 2, \dots m)$ *is the predicate term.* $\Theta = \Sigma \oplus \Lambda = \{t_1, t_2, \dots t_i, \dots t_n\} \oplus \{p_1, p_2, \dots p_j, \dots p_m\}$ *denotes the logical plus operation of sets* Σ *and* Λ.

Thus, several general characteristics of knowledge elements can be introduced from Definition 5: (1) knowledge elements have a certain structure and constitute the smallest controllable unit of process innovation knowledge; (2) knowledge elements are logically complete and capable of expressing facts, principles, methods, and so on; (3) new knowledge can be generated by semantically correlating multi-sourced knowledge elements.

Definition 6. *Natural language description D and its composition proposition* P_i *of process innovation knowledge element can be further represented as*

$$\begin{aligned} D &\triangleq \cup P_i (i = 1, 2, \dots, n), \\ P_i &\triangleq \cup (t_{ij}) \oplus \cup (p_{ij}) \ (i = 1, 2, \dots, n; j = 1, 2, \dots m) \end{aligned} \tag{6}$$

where $t_{ij} (i = 1, 2, \dots, n; j = 1, 2, \dots m)$ *is terminology of proposition, and* $p_{ij} (i = 1, 2, \dots, n; j = 1, 2, \dots m)$ *is the predicate of proposition.*

Knowledge fusion is a process of forming new knowledge, with the help of multi-source knowledge interaction and support. For terminology fusion, the terminology specification and terminology conflict resolution of the fusion process are based on domain ontology and semantic relationships. For knowledge element sets of natural language description, we can deconstruct them as subject–predicate–object (SPO) logical form triples and then reconfigure semantic elements through co-reference relationship identification under the domain ontology constraints.

The algorithm flow of knowledge fusion for process innovation knowledge is represented in Figure 3, and the specific process is given as follows:

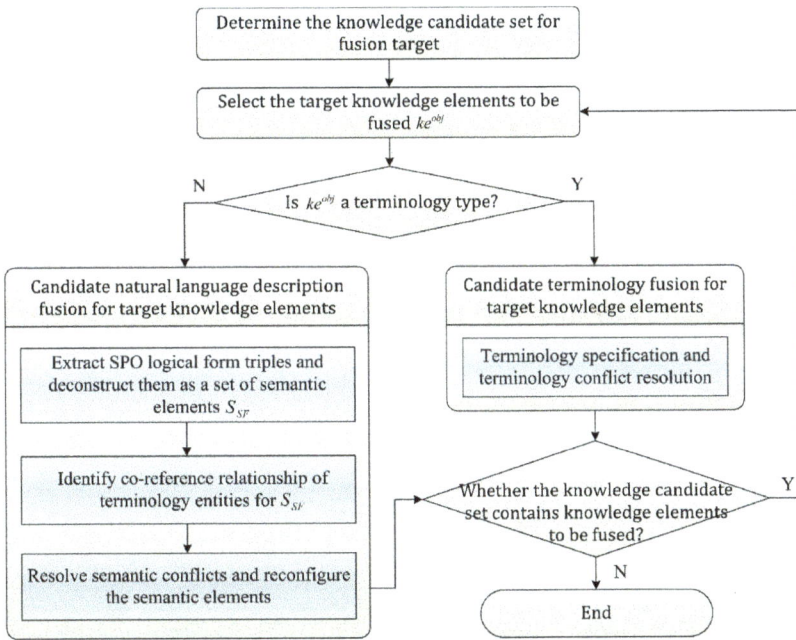

Figure 3. Algorithm flow of process innovation knowledge fusion based on semantic elements reconfiguration.

Step 1. Determine the knowledge candidate set for fusion target of knowledge unit KN_k^{cnd} and knowledge contextual relevance CTR_t^{cnd}.

$$KN_k^{cnd} = \{< P_k, I_{Ik}, I_{Ok}, E_k, u_k > |k = 1, 2, \ldots, n\},$$
$$CTR_t^{cnd} = \{\langle kn_t, k_t, r_t, k'_t, u_t \rangle |t = 1, 2, \ldots, m\} \tag{7}$$

where n and m are the number of knowledge unit candidates and the number of knowledge contextual relevance candidates in the fusion process, respectively.

Step 2. Select the target knowledge elements of knowledge candidate set Ke^{obj}, and judge whether it is a terminology type. If so, then go to Step 3, otherwise turn to Step 4.

Step 3. Standardize the candidate terminology set and perform logical plus operation based on domain ontology. If completed, turn to step 7.

For two knowledge elements Ke_i, Ke_j in fusion process, if there are terminology items $t_i \in Ke_i$, $t_j \in Ke_j$ and terminology conflict $t_i \times t_j$, those conflicts will be resolved according to the following rules:

(1) When the terminologies have similar meanings but different expressions, we can map terminology items t_i, t_j into the terminology set logic tree T of domain ontology, and the result can be denoted by R. If $T(t_i) \subset T(t_j)$, then $R = t_j$; if $T(t_i) \supset T(t_j)$, then $R = t_i$; if $T(t_i) = T(t_j)$, then $R = t_i$ or t_j.

(2) When the terminology items have contrary logic, conflict resolution will depend on collective intelligence.

Step 4. Execute semantic and grammatical analysis for the candidate natural language descriptions, and extract SPO logical form triples by using semantic linguistic tool NLPWin [54], which provides deep syntactic and partial semantic analysis of text, then deconstruct them as a set of semantic elements S_{SF}.

Step 5. Identify co-reference relationship of terminology entities for S_{SF}. Terminology entities refer to the terms or phrases that are defined by the domain ontology, such as the manufacturing resources, processing objects, process methods, and so on.

Step 6. Perform the logical plus operation for deconstructed natural language descriptions of the candidate set, and reconfigure the semantic elements of S_{SF}. If completed, go to step 7.

For two knowledge elements Ke_i and Ke_j in fusion process, if there are semantic items $(t_i \oplus p_i) \in Ke_i$, $(t_j \oplus p_j) \in Ke_j$ and semantic conflict $(t_i \oplus p_i) \otimes (t_j \oplus p_j)$, those conflicts will be resolved according to the following rules:

(1) When concrete manifestation of semantic conflict is terminology conflict, those conflicts can be resolved according to Step 3.

(2) When predicate items have similar meanings but different expressions, we denote the usage frequency of predicate terms p_i, p_j by f_i and f_j, respectively. Similarly, the fusion result is denoted by R. If $f_i \prec f_j$, then $R = p_j$; if $f_i \prec f_j$, then $R = p_j$; if $f_i = f_j$, then $R = p_i$ or p_j.

(3) When predicate items have contrary logic, conflicts resolution will depend on collective intelligence.

Step 7. Judge whether the candidate knowledge sets KN_k^{cnd} and CTR_i^{cnd} still contain knowledge elements that need to be fused. If so, return to Step 2, otherwise end this algorithm.

4.2. Collaborative Innovation Knowledge Refinement Based on Knowledge–Social Trust Degree

Innovation knowledge fusion unit contains the wisdom of the participants' individual knowledge, yet to some extent it is rough or inaccurate and needs to be refined further by experts and authorities. Knowledge refinement is a collaborative editing process of preliminary knowledge by group members with a high knowledge–social trust degree (*KST*). In order to rapidly capture process innovation knowledge and ensure the credibility and orderliness of the knowledge refinement procedure, we regulate group members' knowledge behavior by applying a collaborative editing mechanism.

4.2.1. Credible Groups Construction

In the process of innovation knowledge capturing, knowledge–social members give comments and evaluations on other members' knowledge activities and establish social trust relationships among them. Here, the participants' knowledge–social trust degree in a knowledge community is measured by two aspects: individual trust (KST_{ind}) and community trust (KST_{com}).

Definition 7. *KST_{ind} is used to describe the trust level established on knowledge interaction between one user and another user. Suppose there are individuals d_i and d_j in the knowledge–social community, d_i and d_j had n_1 times knowledge–social activities which has an interactive type of P_h. Let $jud_{d_j}(d_i) \in [0, 1]$ be an interactive evaluation of d_j toward d_i in a knowledge–social activity. Assuming that d_j has given m_1 times negative comment on d_i, the KST_{ind} of d_j toward d_i can be computed as:*

$$KST_{ind}(d_j, d_i) = \frac{\sum_{t=1}^{n_1} right(P_h) \times jud_{d_j}(d_i)}{n_1} \times \left(\frac{n_1 - m_1}{n_1}\right)^{\frac{1}{n_1 - m_1}} \tag{8}$$

where $right(P_h) \in [0, 1]$ is weight coefficient of interactive type. This formula introduces the weight concept of knowledge interaction and considers the influence of malicious interaction on subjective trust, which makes the calculation more reliable.

Definition 8. *KST_{com} indicates the overall trust and reliability of users in the knowledge–social community, given by all members of the community in which the individual resides. The KST_{com} calculation depends on the following two factors: (1) the common evaluation for someone's knowledge–social behavior from all members of knowledge community; (2) the number of knowledge communities in which this individual resides. Suppose there* *is an individual* $d_i \in V$ *in multiple knowledge communities* $V_1, V_2, \ldots, V_e$. *Assuming that* d_i *has been evaluated by* g *members of knowledge communities* $V_1, V_2, \ldots, V_e$, *we can obtain the* KST_{com} *of* d_i.

$$KST_{com}(d_i) = 1 \bigg/ g \times \sum_{\substack{d_j \in V_1 \cup V_2 \cup \ldots \cup V_e \\ j \neq i}} \left[KST_{ind}(d_j, d_i)^{\frac{1}{|V_{d_j}|}} \times (KST_{com}(d_j)) \right], \qquad (9)$$

where V *is the knowledge community set, and* $\left| V_{d_j} \right|$ *is the number of knowledge communities* $V_1, V_2, \ldots, V_e$ *in which the individual* d_j *resides. Considering the extensive influence of community participants, the number of communities is introduced as a factor in* KST_{com} *calculation. If a participant has identities in multiple knowledge communities, the influence from his evaluation will be more than the one from only one community. In the process of knowledge refinement, the credibility of knowledge refined by participants with multiple identities will certainly be higher than that refined by the user with single community identity.*

Suppose there are t members in a group G, the degrees of group knowledge-social trust $KST_{com}(d_i)$, $KST_{com}(d_j)$ have not been determined. The specific procedures of credible groups construction based on KST are summarized as follows:

Step 1. Compute individual knowledge–social trust degree $KST_{ind}(d_j, d_i)$ for t members of group G by using Formula (8).

Step 2. Initialize community knowledge–social trust degree for each group member i, $KST_{com}(d_i) = k \in (0, 1]$.

Step 3. Calculate temporary community knowledge–social trust degree $\overline{KST_{com}(d_i)}$ of each group member by applying Formula (9):

$$\overline{KST_{com}(d_i)} = 1 \bigg/ g \times \sum_{\substack{d_j \in V_1 \cup V_2 \cup \ldots \cup V_e \\ j \neq i}} \left[KST_{ind}(d_j, d_i)^{\frac{1}{|V_{d_j}|}} \times (KST_{com}(d_j)) \right] \qquad (10)$$

Step 4. Judge whether the KST_{com} satisfies accuracy error according to the following formula:

$$\sum \left| \overline{KST_{com}(d_i)} - KST_{com}(d_i) \right| < \Delta \qquad (11)$$

where Δ is the setting accuracy error value. If so, go to Step 5; otherwise let $KST_{com}(d_i) = \overline{KST_{com}(d_i)}$ for each knowledge–social member, and return to Step 3.

Step 5. Structure the KST_{com} set of knowledge–social members, $KST_{com} = \{KST_{com}(d_i) | i \in V\}$. Select the members with higher KST to join the knowledge refinement group based on the following basic criterion:

$$KST_{com}(d_i) \geq \xi \qquad (12)$$

where ξ is the knowledge–social trust threshold, which can be set based on the requirements of innovation knowledge refinement.

4.2.2. Procedure of Collaborative Knowledge Refinement

Knowledge refinement process requires collective participation of knowledge–social users, and refinement results should include ideas from knowledge refinement members as much as possible. In an open knowledge–social community, members of credible knowledge group have the permission of the corresponding knowledge editing and refinement. The procedure of collaborative process innovation knowledge refinement is displayed in Figure 4. Firstly, managers propose knowledge refinement requirements and build refinement groups according to the knowledge to be refined.

Then group members discuss original knowledge object K_0, publish their suggestions for revision and post their attitudes toward the views of others. A suitable member u_1 will be selected as the knowledge editor to perform refinement transaction. Thus, a temporary knowledge version K_1^1 is formed by the first-round editor u_1. When knowledge editing of this round is completed, the members make an editorial comment on version K_1^1 again and carry out the procedure of knowledge refinement. Then repeat the above process until the knowledge is fully refined. As shown in Figure 4, through the gradual refinement for original knowledge object K_0 by editors $u_1 \ldots u_n$, multiple temporary versions may have been correspondingly formed as knowledge versions $K_1^1 \ldots K_1^n$. When the latest temporary version K_1^n reach the refinement requirements, it will be saved as the refinement result of this time K_1. In addition, with the knowledge application in manufacturing process innovation design, the new requirement of knowledge refinement will still be put forward.

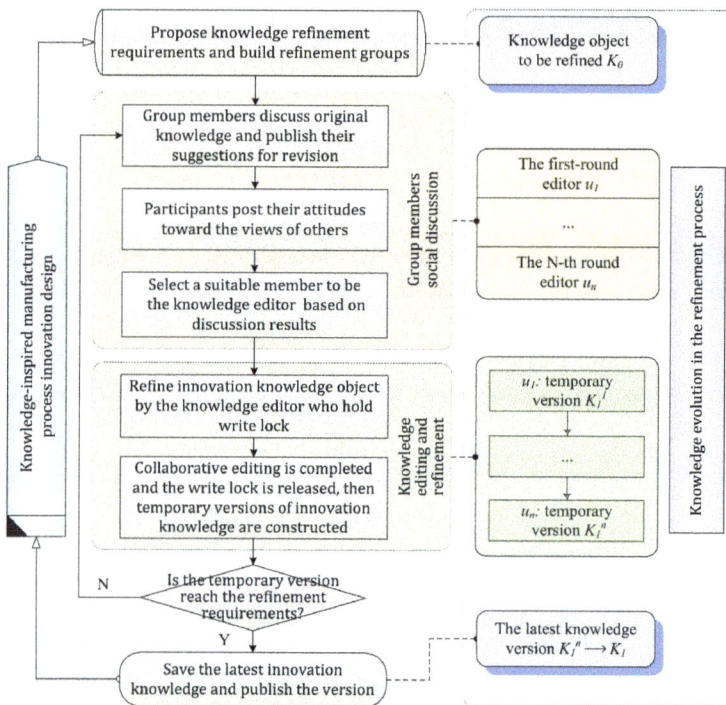

Figure 4. Procedure of collaborative knowledge refinement in an open knowledge–social community.

While accessing any knowledge elements of collaborative editing, group members can take the following actions: view and edit the existing knowledge. Because multiple users may execute transactions simultaneously in the refinement process and the transactions are atomic, knowledge element modification for different transactions should be mutually exclusive. Here, a lock-based knowledge collaborative editing and refinement solution is adopted to enable concurrent access to workflows for multiple knowledge editors, and its specific rules are shown in Table 1. Knowledge locks, in this study, basically consist of two types: read locks and write locks. The editor who owns the write lock has editing permission for the locked region, while the read lock owner is only allowed to read knowledge content. To avoid redundant effort and to prevent editors from destroying each other's work, the write locks are exclusive in this research.

Table 1. Rules for lock-based knowledge collaborative editing and refinement (adapted from [25]).

Rules	Descriptions
Rule 1	The read locks are compatible with each other. More than one read locks can be placed on one knowledge object at the same time. Group members of knowledge refinement are allowed to hold read lock of the corresponding knowledge objects.
Rule 2	The write locks are mutually exclusive with each other for a locked region. This means that only one write lock can be placed on the same knowledge object at a certain moment, and for a knowledge element only one editor may hold the write lock.
Rule 3	After participants publish their comments and exchange views on the knowledge object to be refined, members who obtained a positive evaluation of more than a certain level can apply to be the refinement editor.
Rule 4	If a knowledge element has been locked, the write lock requests will be put forward. Meanwhile, notifications are sent to the owners of write locks whenever the latter form queues in front of certain knowledge objects. Specifically, a system timer process, which sends time-stamped notifications to the owners of write locks, can be employed to prevent the starvation of other editing operations whenever there are editing operations waiting for more than a certain time to access certain objects.
Rule 5	Group managers have the permission to grant write locks to a suitable group member at all times.

5. Case Study

5.1. The Implementation of MPI-OKCS

Based on the proposed approach, this sub-section implements a prototype management system *MPI-OKCS* for open capturing systematic process innovation knowledge. It is integrated as a submodule into the pre-built general platform of CAPI system *pi*Pioneer, which contains the basic tools needed for the knowledge management system.

The *MPI-OKCS* has a 4-layer-architecture, as illustrated in Figure 5. The knowledge & data layer stores the basic data of the innovation system, knowledge–social information of the community, and captured process innovation knowledge. The service layer supports access to the knowledge and data layer, and provides various system background services of knowledge capturing process. The functional layer provides the functional components required for the system business logic of the three main modules, namely, knowledge capture, knowledge application, and system management. The interaction layer provides a visual man–machine interface for users from different departments and dispersed geographic locations, so that they can participate in the innovation knowledge processing activities of the corresponding roles in an open environment.

To facilitate the implementation process, we have invited seven domain experts from the Institute of CAPP & Manufacturing Engineering Software at NWPU (Xi'an, China) and the Department of Mechanical Design at CHD (Xi'an, China) to participate in innovation knowledge refinement. All graduate students from the above two departments were allowed to contribute their innovation knowledge. Additionally, about 20 engineers from the R & D department of Sinomach Changlin Company Limited (Changzhou, China) have contributed their individual process knowledge to the system.

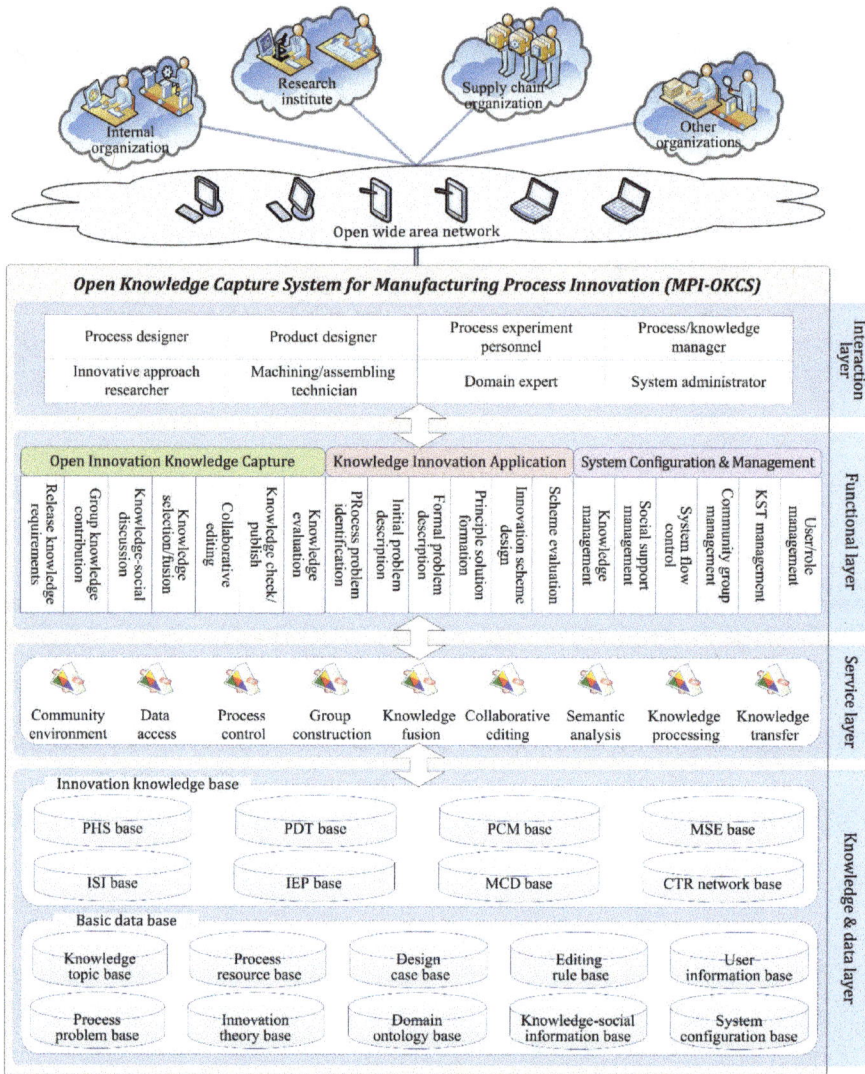

Figure 5. The system architecture of *MPI-OKCS*.

5.2. An Illustrative Example of Welding Process Innovation Knowledge Capture and Application

Welding technology is widely used in the manufacture of aerospace vehicles, electronic precision instruments, pressure vessels and so on. With the complexity and diversification of product requirements, the specific process issues to be solved in welding technology are also increasing. In the following, we take welding process innovation as an example to illustrate the concrete process of open innovation knowledge capture.

Figure 6 presents the procedures of knowledge capture for the circuit board welding process problem-solving of an electronic device. Firstly, the system publishes knowledge topics and problem-solving requirements, then notifies the related knowledge–social users. According to the situation of process problem solving, multiple types of innovation knowledge can be included: PCM,

MSE, ISI, et al. Here, the knowledge type of process contradiction is selected as required in this round of knowledge capturing (as shown in Part 1 of Figure 6).

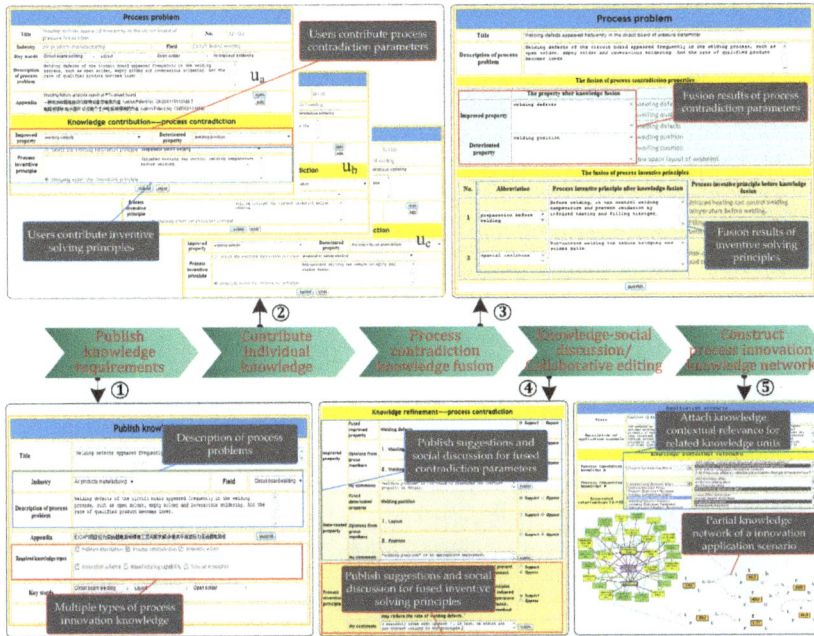

Figure 6. An instance of welding process innovation knowledge capture process. Part 1: Publish knowledge requirements; Part 2: Contribute individual knowledge; Part 3: Process contradiction knowledge fusion; Part 4: Knowledge-social discussion/Collaborative editing; Part 5: Construct process innovation knowledge network.

Those interested users are formed into the knowledge contribution group, then they discuss the knowledge topics and exchange views, and contribute their individual knowledge according to the corresponding knowledge templates. In Part 2, three members u_a, u_b and u_c have respectively contributed their process contradiction knowledge, which contain contradiction parameters and corresponding inventive solving principles. Three pairs of process contradiction parameters are as follows: P_a = <*welding defects → welding position*>, P_b = <*welding quality → welding position*> and P_c = <*welding defects →* the space layout of weldment>. And three natural language descriptions of inventive solving principles are as follows: D_a = {*Infrared heating can control welding temperature before welding*}, D_b = {*Filling nitrogen can prevent oxidation before welding*}, and D_c = {*Non-contact welding can reduce bridging and solder balls*}.

Subsequently, the above process contradiction knowledge is further fused together, as illustrated in Part 3 of Figure 6. According to the relationships of process terms ontology, three process parameters to be improved are fused into a result for the strengthening process parameter, *welding defects*. Similarly, the fusion result of the weakening process parameter, *welding position*, is obtained. Thus, the fused process contradiction parameters can be expressed as P_F = <*welding defects → welding position*>. Meanwhile, the system will extract the logical form triples of three innovation principle descriptions. From the extraction results in Figure 7, P_a and P_b have the specific semantic association, and they can form a fused semantic graph. Furthermore, with the support of process resources and knowledge of the general platform *pi*Pioneer, a fusion result of innovation principle descriptions can be formed using the semantic elements reconfiguration method. The fusion results are described as follows:

D_F = {*By combining the use of infrared heating and filling nitrogen before welding, the welding temperature can be effectively controlled and the oxidation can be prevented*}. Figure 7 shows the fusion process of process innovation principle descriptions.

In a knowledge–social community, the preliminary fused process contradiction knowledge will be transferred to the credible knowledge refinement group formed with high *KST* members. Refinement members can publish revision suggestions and have a chance to get the write lock. Through knowledge–social discussion and multiple rounds of collaborative editing, the refined welding process contradiction knowledge for this knowledge topic/problem-solving is captured. In the same way, the capture procedures of other knowledge types are basically consistent with process contradiction.

Parts 1–4 of Figure 6 give the description for innovation knowledge capture of PCM type. Similarly, other types of process innovation knowledge units can also be captured by this way. When the number of process innovation knowledge units is sufficient, knowledge contextual relevance can be attached to the related units to form a specific knowledge network, which has a certain problem-solving ability in the semantic environment. Based on the published application scenario, knowledge–social users can contribute their individual knowledge contextual relevance by selecting knowledge types, knowledge entries, and the corresponding associated relationships, as shown in Part 5 of Figure 6. Correspondingly, innovation knowledge network construction for a specific innovation application scenario needs not only a large number of multi-type knowledge units, but also the new round of knowledge–social members' collaborative editing based on collective intelligence. In this case study, after about six months of open knowledge capture and welding knowledge accumulation in the pre-research stage, an innovation knowledge network for problem solving of circuit board welding was built in the *MPI-OKCS*. Part 5 of Figure 6 gives a partial knowledge network for the above innovation application scenario, which currently contains 223 refined knowledge units. Among them, a welding process contradiction matrix is captured, as illustrated in Figure A1 and Tables A1 and A2. With the aid of the innovation application module of *pi*Pioneer, the captured innovation knowledge units and knowledge networks have played an effective role in inspiring the process problem-solving for development of a new-type pressure sensor.

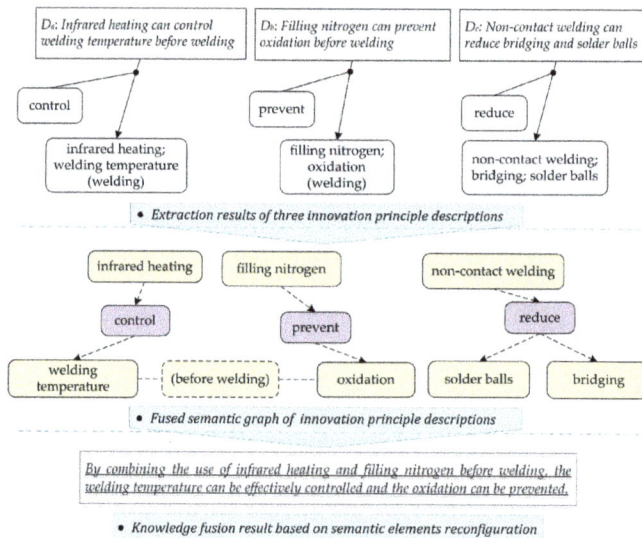

Figure 7. A fusion example of process innovation principle descriptions.

6. Conclusions and Implications

Manufacturing process innovation has been recognized as a key factor for reducing production costs, improving product quality, and enhancing sustainable competitive edge. Nevertheless, in the implementation of knowledge-driven CAPI, an important challenge that must be faced is how to effectively capture the structured, formalized, and associated innovation knowledge from empirical knowledge owners. In this paper, we have presented an integrated approach for processing innovation knowledge capture based on collective intelligence. Some of the main contributions of this research are listed below:

- By considering the multi-type knowledge organization in innovation design and building a knowledge–social community, a novel holistic knowledge capture paradigm of PIKC-CI is proposed, which can realize the transformation from individual empirical knowledge to public refined knowledge in an open environment.
- Based on the domain ontology constraints, a multi-source process innovation knowledge fusion algorithm based on semantic elements reconfiguration is raised, with the corresponding semantic conflict resolution rules. This algorithm can effectively support preliminary automatic fusion for the contributed knowledge.
- A collaborative editing strategy based on knowledge lock and *KST* is applied to the iterative refinement of process innovation knowledge, which ensures that refined knowledge embraces the collective intelligence of knowledge–social users.

Potential future studies related to this work are as follows. Firstly, in addition to the current static knowledge network for specific application scenarios, we are interested in studying how to construct the innovation problem-oriented dynamic knowledge network. Secondly, we will expand our approach to the automatic knowledge capture from problem-solving schemes of the process planning system, and manufacturing process-related text of the cloud manufacturing platform. Moreover, from the perspective of knowledge application, it is worth exploring how to realize just-in-time knowledge recommendations for innovation design life cycle.

Acknowledgments: This work was supported by the Fundamental Research Funds for the Central Universities of China (Grant Nos. 310825171004 and 310825173314). In addition, G.W. would also like to express special gratitude to CAPP team of NWPU for their support in CAPI project research.

Author Contributions: G.W., X.T., and J.G. conceived this study; G.W. and Y.H. drafted the manuscript and improved the knowledge processing algorithms; G.H. and M.Z. proofread and revised the content of the original manuscript. All authors have read and approved the final manuscript.

Conflicts of Interest: The authors declare no conflict of interest.

Appendix A

See Figure A1 and Tables A1 and A2.

		The weakening process parameters												
		1	2	3	4	5	6	7	8	9	10	11	12	...
The strengthening process parameters	1			<7>							<7>	<7>		
	2	<7,9,10>	<8>	<3>			<6>							
	3											<7>		
	4	<7>						<1,4>						
	5		<6,3,2,10>		<1,2,8>									
	6			<3,6>										
	7				<4>									
	8													
	9	<7>		<10>				<6>			<9,10>			
	10	<9>	<9>										<3,8>	
	11				<7>	<2>	<6>	<2,8>	<5>	<7>	<9>			
	12		<3,8>								<3,8>			
	...													

Figure A1. Welding process contradiction matrix.

Table A1. Contradiction parameters of welding process.

No.	Parameters	Explanations
1	Material	Physical and chemical properties of materials
2	Mechanical properties	Stress, pressure, tensile strength etc.
3	Thickness	Thickness range of different materials can be welded
4	Strength	Mechanical strength after welding
5	Shape	Break/joint form, welding wire size, weld shape/aspect ratio, arc spacing etc.
6	Welding position	Butt contact, angular contact, lap joint, downward welding, vertical, horizontal and inverted welding, constraint degree
7	Temperature	Preheat temperature, heat treatment temperature, cooling temperature, temperature distribution etc.
8	Power	Welding current, arc voltage, power supply
9	Speed	Welding speed, wire feed speed, wire melting speed, cooling rate etc.
10	Oxidability	Heat input, weld/base metal oxidation
11	Welding defects	Appearance defects, surface defects, cracks, incomplete penetration, not fusion etc.
12	Production efficiency	Welding utilization, product efficiency
...	...	...

Table A2. Contradiction solving principles of welding process.

No.	Principles	Explanations
1	Separation/detachment/compromise	a. Divide objects into separate parts; b. Make the object detachable; c. Increase the object segmentation.
2	Preparation before welding	a. The choice and treatment of the crevasses form; b. Pre-calculation processing.
3	Change one-dimension to multi-dimension (new dimension)	a. The material motion in the form of point, one-dimensional, two-dimensional, three-dimensional spatial distribution or conversion; b. Replacing single layer structure with multi-layer structure; c. Incline, side, or invert the object; d. To the opposite or adjacent surface of a specified surface.
4	Heat treatment	a. Normalizing; b. Quenching; c. Tempering; d. Annealing.
5	Turn the harm into benefit	a. Use harmful factors (especially the harmful effects of the medium) to gain beneficial effects; b. Harmful factors can be eliminated by a combination of harmful factors and one or more other harmful factors; c. Improve the extent of the operation of the harmful factors in order to achieve a state of harmless.

Table A2. *Cont.*

No.	Principles	Explanations
6	Substitution/replacement principle	a. Using two or more welding methods instead of a single welding method; b. The new welding consumables and solder are used to replace the old ones; c. The quantitative and faintness factors, fixed and variable parameters, irregular and regular state are converted into each other in welding; d. Using high energy density energy.
7	Welding material selection	a. Select stainless steel consumables according to ASME specifications; b. Select welding consumables by application or composition.
8	Dispersion principle (homogeneity)	a. The welding consumables should be of the same material (or of the similar mechanical properties) when welding a given object. B. Distract the stress of the stress concentration part.
9	Setting media protection	a. Replacing the normal environment with an inert environment; b. Introduction of a mixture or additive; c. Welding process in vacuum environment.
10	Composite/hybrid principle	a. Transfer from the same material to the mixture; b. Substitute a composition for a similar substance.
...	...	...

Abbreviations

The following abbreviations are used in this manuscript:

CAPI	computer-aided process innovation
CAM	computer-aided manufacturing
CAPP	computer-aided process planning
CAD	computer-aided design
CAE	computer-aided engineering
CAI	computer-aided innovation
ICT	information and communication technology
KC	knowledge contribution
KF	knowledge fusion
KR	knowledge refinement
TRIZ	the theory of inventive problem solving
PIKC-CI	process innovation knowledge capture based on collective intelligence
PHS	problem heuristic scene
PDT	problem description template
PCM	process contradiction matrix
MSE	manufacturing scientific effect
ISI	innovative scheme instance
IEP	innovative evaluation parameter
MCD	manufacturing capability description
KST	knowledge–social trust degree
MPI-OKCS	open knowledge capture system for manufacturing process innovation

References

1. Neugebauer, R. Energy-Efficient Product and Process Innovations in Production Engineering. *CIRP J. Manuf. Sci. Technol.* **2011**, *4*, 127–128. [CrossRef]
2. Wang, G.; Tian, X.; Geng, J.; Guo, B. A Process Innovation Knowledge Management Framework and Its Application. *Adv. Mater. Res.* **2013**, *655–657*, 2299–2306. [CrossRef]
3. Geng, J.; Tian, X.; Jia, X.; Liu, S.; Zhang, Z. Review for computer aided methods of manufacturing process innovation. *Comput. Integr. Manuf. Syst.* **2016**, *22*, 2778–2790.
4. Jafari, M.; Zarghami, H.R. Effect of TRIZ on enhancing employees' creativity and innovation. *Aircraft Eng. Aerosp. Technol.* **2017**, *89*, 853–861. [CrossRef]

5. Sun, X.; Shehab, E.; Mehnen, J. Knowledge modelling for laser beam welding in the aircraft industry. *Int. J. Adv. Manuf. Technol.* **2012**, *66*, 763–774. [CrossRef]
6. Wang, C.N.; Huang, Y.F.; Le, T.N.; Ta, T.T. An innovative approach to enhancing the sustainable development of Japanese automobile suppliers. *Sustainability* **2016**, *8*, 420. [CrossRef]
7. Yi, P.X.; Huang, M.; Guo, L.J.; Shi, T.L. Dual recycling channel decision in retailer oriented closed-loop supply chain for construction machinery remanufacturing. *J. Clean. Prod.* **2016**, *137*, 1393–1405. [CrossRef]
8. Kang, D.; Lee, D.H. Energy and environment efficiency of industry and its productivity effect. *J. Clean. Prod.* **2016**, *135*, 184–193. [CrossRef]
9. May, G.; Stahl, B.; Taisch, M.; Kiritsis, D. Energy management in manufacturing: From literature review to a conceptual framework. *J. Clean. Prod.* **2017**, *167*, 1464–1489. [CrossRef]
10. Kobayashi, H. A systematic approach to eco-innovative product design based on life cycle planning. *Adv. Eng. Inform.* **2006**, *20*, 113–125. [CrossRef]
11. U.S. President's Council of Advisors on Science and Technology. *Report to the President on Capturing Domestic Competitive Advantage in Advanced Manufacturing*; U.S. President's Council of Advisors on Science and Technology: Washington, DC, USA, 2012.
12. Ramos, T.B.; Martins, I.P.; Martinho, A.P.; Douglas, C.H.; Painho, M.; Caeiro, S. An open participatory conceptual framework to support State of the Environment and Sustainability Reports. *J. Clean. Prod.* **2014**, *64*, 158–172. [CrossRef]
13. Alvarado, A. Problems in the implementation process of advanced manufacturing technologies. *Int. J. Adv. Manuf. Technol.* **2013**, *64*, 123–131.
14. Hollen, R.M.A.; Van den Bosch, F.A.J.; Volberda, H.W. The Role of Management Innovation in Enabling Technological Process Innovation: An Inter-Organizational Perspective. *Eur. Manag. Rev.* **2013**, *10*, 35–50. [CrossRef]
15. Mani, M.; Madan, J.; Lee, J.H.; Lyons, K.W.; Gupta, S.K. Sustainability characterisation for manufacturing processes. *Int. J. Prod. Res.* **2014**, *52*, 5895–5912. [CrossRef]
16. Shin, S.J.; Kim, D.B.; Shao, G.; Brodsky, A.; Lechevalier, D. Developing a decision support system for improving sustainability performance of manufacturing processes. *J. Intell. Manuf.* **2017**, *28*, 1421–1440. [CrossRef]
17. Främling, K.; Holmström, J.; Loukkola, J.; Nyman, J.; Kaustell, A. Sustainable PLM through Intelligent Products. *Eng. Appl. Artif. Intell.* **2013**, *26*, 789–799. [CrossRef]
18. Liu, L.; Jiang, Z.; Song, B. A novel two-stage method for acquiring engineering-oriented empirical tacit knowledge. *Int. J. Prod. Res.* **2014**, *52*, 5997–6018. [CrossRef]
19. Stadler, C. Process Innovation and Integration in Process-Oriented Settings: The Case of the Oil Industry. *J. Prod. Innov. Manag.* **2011**, *28*, 44–62. [CrossRef]
20. Wang, G.; Tian, X.; Hu, Y.; Evans, R.D.; Tian, M.; Wang, R. Manufacturing Process Innovation-Oriented Knowledge Evaluation Using MCDM and Fuzzy Linguistic Computing in an Open Innovation Environment. *Sustainability* **2017**, *9*, 1630. [CrossRef]
21. Alexander, A.T.; Childe, S.J. Innovation: A knowledge transfer perspective. *Prod. Plan. Control* **2013**, *24*, 208–225. [CrossRef]
22. Gao, J.; Bernard, A. An overview of knowledge sharing in new product development. *Int. J. Adv. Manuf. Technol.* **2017**, *94*, 1545–1550. [CrossRef]
23. Leo Kumar, S.P. Knowledge-based expert system in manufacturing planning: State-of-the-art review. *Int. J. Prod. Res.* **2018**, 1–25. [CrossRef]
24. Waris, M.M.; Sanin, C.; Szczerbicki, E. Smart Innovation Engineering: Toward Intelligent Industries of the Future. *Cybern. Syst.* **2018**, 1–16. [CrossRef]
25. Wang, G.; Tian, X.; Geng, J.; Guo, B. A knowledge accumulation approach based on bilayer social wiki network for computer-aided process innovation. *Int. J. Prod. Res.* **2015**, *53*, 2365–2382. [CrossRef]
26. Woolley, A.W.; Chabris, C.F.; Pentland, A.; Hashmi, N.; Malone, T.W. Evidence for a collective intelligence factor in the performance of human groups. *Science* **2010**, *330*, 686–688. [CrossRef] [PubMed]
27. Denkena, B.; Shpitalni, M.; Kowalski, P.; Molcho, G.; Zipori, Y. Knowledge Management in Process Planning. *CIRP Ann. Manuf. Technol.* **2007**, *56*, 175–180. [CrossRef]
28. Quintane, E.; Mitch Casselman, R.; Sebastian Reiche, B.; Nylund, P.A. Innovation as a knowledge-based outcome. *J. Knowl. Manag.* **2011**, *15*, 928–947. [CrossRef]

29. Esterhuizen, D.; Schutte, C.S.L.; du Toit, A.S.A. Knowledge creation processes as critical enablers for innovation. *Int. J. Inf. Manag.* **2012**, *32*, 354–364. [CrossRef]

30. Xu, J.; Houssin, R.; Caillaud, E.; Gardoni, M. Fostering continuous innovation in design with an integrated knowledge management approach. *Comput. Ind.* **2011**, *62*, 423–436. [CrossRef]

31. Bosch-Mauchand, M.; Belkadi, F.; Bricogne, M.; Eynard, B. Knowledge-based assessment of manufacturing process performance: Integration of product lifecycle management and value-chain simulation approaches. *Int. J. Comput. Integr. Manuf.* **2013**, *26*, 453–473. [CrossRef]

32. Wang, G.; Tian, X.; Geng, J.; Evans, R.; Che, S. Extraction of Principle Knowledge from Process Patents for Manufacturing Process Innovation. *Proc. CIRP* **2016**, *56*, 193–198. [CrossRef]

33. Van de Vrande, V.; de Jong, J.P.J.; Vanhaverbeke, W.; de Rochemont, M. Open innovation in SMEs: Trends, motives and management challenges. *Technovation* **2009**, *29*, 423–437. [CrossRef]

34. Huizingh, E.K.R.E. Open innovation: State of the art and future perspectives. *Technovation* **2011**, *31*, 2–9. [CrossRef]

35. Carbone, F.; Contreras, J.; Hernández, J.Z.; Gomez-Perez, J.M. Open Innovation in an Enterprise 3.0 Framework: Three Case Studies. *Expert Syst. Appl.* **2012**, *39*, 8929–8939. [CrossRef]

36. Cappa, F.; Del Sette, F.; Hayes, D.; Rosso, F. How to deliver open sustainable innovation: An integrated approach for a sustainable marketable product. *Sustainability* **2016**, *8*, 1341. [CrossRef]

37. Hüsig, S.; Kohn, S. "Open CAI 2.0"—Computer Aided Innovation in the era of open innovation and Web 2.0. *Comput. Ind.* **2011**, *62*, 407–413. [CrossRef]

38. Schumpeter, J.A. *The Theory of Economic Development*; Harvard University Press: Cambridge, MA, USA, 1934.

39. Ayhan, M.B.; Öztemel, E.; Aydin, M.E.; Yue, Y. A quantitative approach for measuring process innovation: A case study in a manufacturing company. *Int. J. Prod. Res.* **2013**, *51*, 3463–3475. [CrossRef]

40. Leon, N.; Cho, S.K. Computer aided innovation. *Comput. Ind.* **2009**, *60*, 537–538. [CrossRef]

41. Kiritsis, D.; Koukias, A.; Nadoveza, D. ICT supported lifecycle thinking and information integration for sustainable manufacturing. *Int. J. Sustain. Manuf.* **2014**, *3*, 229–249. [CrossRef]

42. Ilevbare, I.M.; Probert, D.; Phaal, R. A review of TRIZ, and its benefits and challenges in practice. *Technovation* **2013**, *33*, 30–37. [CrossRef]

43. Xu, X.; Wang, L.; Newman, S.T. Computer-aided process planning—A critical review of recent developments and future trends. *Int. J. Comput. Integr. Manuf.* **2011**, *24*, 1–31. [CrossRef]

44. Lukic, D.; Milosevic, M.; Antic, A.; Borojevic, S.; Ficko, M. Multi-criteria selection of manufacturing processes in the conceptual process planning. *Adv. Prod. Eng. Manag.* **2017**, *12*, 151–162. [CrossRef]

45. Raymond, L.; Bergeron, F.; Croteau, A.-M. Innovation capability and performance of manufacturing SMEs: The paradoxical effect of IT integration. *J. Organ. Comput. Electron. Commer.* **2013**, *23*, 249–272. [CrossRef]

46. Yusof, Y.; Latif, K. Survey on computer-aided process planning. *Int. J. Adv. Manuf. Technol.* **2014**, *75*, 77–89. [CrossRef]

47. Cakir, M.C.; Cilsal, O.O. Implementation of a contradiction-based approach to DFM. *Int. J. Comput. Integr. Manuf.* **2008**, *21*, 839–847. [CrossRef]

48. Duflou, J.R.; D'hondt, J. Applying TRIZ for systematic manufacturing process innovation: The single point incremental forming case. *Proc. Eng.* **2011**, *9*, 528–537. [CrossRef]

49. Sheu, D.D.; Chen, C.-H.; Yu, P.-Y. Invention principles and contradiction matrix for semiconductor manufacturing industry: Chemical mechanical polishing. *J. Intell. Manuf.* **2012**, *23*, 1637–1648. [CrossRef]

50. Geng, J.; Tian, X. Knowledge-Based Computer Aided Process Innovation Method. *Adv. Mater. Res.* **2010**, *97–101*, 3299–3302. [CrossRef]

51. Guo, B.; Geng, J.; Wang, G. Knowledge Fusion Method of Process Contradiction Units for Process Innovation. *Proc. Eng.* **2015**, *131*, 816–822. [CrossRef]

52. Duran-Novoa, R.; Leon-Rovira, N.; Aguayo-Tellez, H.; Said, D. Inventive problem solving based on dialectical negation, using evolutionary algorithms and TRIZ heuristics. *Comput. Ind.* **2011**, *62*, 437–445. [CrossRef]

53. Lykourentzou, I.; Papadaki, K.; Vergados, D.J.; Polemi, D.; Loumos, V. CorpWiki: A self-regulating wiki to promote corporate collective intelligence through expert peer matching. *Inf. Sci.* **2010**, *180*, 18–38. [CrossRef]
54. Heidorn, G. Intelligent Writing Assistance. In *Handbook of Natural Language Processing*; CRC Press: Boca Raton, FL, USA, 2000; pp. 181–207.

applied
sciences

MDPI

Article

A Modified Method for Evaluating Sustainable Transport Solutions Based on AHP and Dempster–Shafer Evidence Theory

Luyuan Chen [1] and Xinyang Deng [2,*]

[1] School of Computer, Northwestern Polytechnical University, Xi'an 710072, China;
   chenluyuan@mail.nwpu.edu.cn
[2] School of Electronics and Information, Northwestern Polytechnical University, Xi'an 710072, China
* Correspondence: xinyang.deng@nwpu.edu.cn; Tel.: +86-29-8843-1267

Received: 25 January 2018; Accepted: 3 April 2018; Published: 5 April 2018

Abstract: With the challenge of transportation environment, a large amount of attention is paid to sustainable mobility worldwide, thus bringing the problem of the evaluation of sustainable transport solutions. In this paper, a modified method based on analytical hierarchy process (AHP) and Dempster–Shafer evidence theory (D-S theory) is proposed for evaluating the impact of transport measures on city sustainability. AHP is adapted to determine the weight of sustainability criteria while D-S theory is used for data fusion of the sustainability assessment. A Transport Sustainability Index (TSI) is presented as a primary measure to determine whether transport solutions have a positive impact on city sustainability. A case study of car-sharing is illustrated to show the efficiency of our proposed method. Our modified method has two desirable properties. One is that the BPA is generated with a new modification framework of evaluation levels, which can flexibly manage uncertain information. The other is that the modified method has excellent performance in sensitivity analysis.

Keywords: sustainability evaluation; analytic hierarchy process; Dempster–Shafer evidence theory; Transport Sustainability Index; car-sharing; sensitivity analysis

---

1. Introduction

Transport has become the basis for the daily operation of society and economy, yet the reliance on transportation as a daily function is a substantive contributor to numerous problems faced by modern society, such as air pollution, noise, congestion, safety, travel delays, and many more [1,2]. To curb these growing problems, sustainable transport has entered the research field of transportation experts and has gradually gained increasing attention [3–8]. Sustainable transport is defined as "transportation that meets mobility needs while also preserving and enhancing human and ecosystem health, economic progress and social justice now and in the future [9]". In other words, sustainable transport needs to promote sustainability in terms of society, environment, and economy. Research is under way to develop sustainable transport solutions, aiming to improve urban transport conditions either in terms of the environment, societal benefits, or the economy.

Instead, our attention is shifted to the evaluation of sustainable transport solutions—especially environment-friendly measures—as they influence city sustainability. These transport solutions include mode sharing such as car-sharing, bike-sharing [10], and park-and-ride systems [11]; intelligent transport solutions like electrical vehicles [12] and plug-in electric vehicles (PEVs) [13]; as well as multi-modal transport solutions [14–16], etc. A broad range of methods and techniques have been proposed to assess the impact of transport solutions on city sustainability. Jeon and Amekudzi [17] developed and determined indicator systems for measuring sustainability in transportation systems.

Litman and Burwell [18] addressed the issues related to the definition, evaluation, and implementation of sustainable transportation. Forty-two techniques that could be used to evaluate the sustainability of urban transportation and 20 commentaries on the mentioned techniques were presented by Wellar [19] in the Transport Canada project.

Recently, Anjali Awasthi et al. [20] put forward a hybrid approach based on analytical hierarchy process (AHP) and Dempster–Shafer evidence theory (D-S theory) to evaluate the influence of environment-friendly transport measures on city sustainability. AHP was firstly used to structure and weight the criteria related to sustainability assessment. Then, the data from multiple information sources was combined using D-S theory and the utility estimation could be obtained. Finally, Transportation Sustainability Indexes (TSIs) with respect to the pre-test stage and post-test stage were compared to make a decision regarding whether the transportation measure had a positive impact on city sustainability and thus could be recommended for adoption in the city. The main advantage of this approach lies in the application of D-S theory to deal with uncertain and incomplete information.

However, the most important issue is not well addressed in Anjali Awasthi et al.'s method [20]: the representation of confidence in evaluation levels for utility is not basic probability assignments (BPAs), but only the probability function [21], which does not make full use of the feature of D-S theory. To address this issue, we propose a modified evidential model for evaluating sustainable transport solutions. The modified method can handle the referred issue in a simple but efficient way, which has two desirable properties: one is that the BPA is generated with a new modification framework of evaluation levels and hence it can flexibly manage uncertain information. The other is that the modified method has excellent performance in sensitivity analysis, which is conducted in terms of the conflictive information of the sustainability assessment and the weight changes of different criteria. Two techniques (AHP and D-S theory) are also referred, as AHP is an effective tool to determine the weight of different criteria in a multi-criteria decision-making (MCDM) problem and D-S theory can manage uncertain, ignorant, and missing information well that are very likely to happen in realistic situations [22–29].

The organization of the rest of this paper is as follows. Section 2 starts with a brief presentation of basic concepts. The proposed evidential model based on AHP and D-S theory to evaluate sustainable transport solutions is introduced in Section 3. Section 4 investigates the implementation of car-sharing in our proposed method. A discussion is presented in Section 5. In Section 6, the paper is ended with a brief summary.

2. Preliminaries

2.1. Dempster–Shafer Theory

Information in the real world is affected by a great deal of uncertainty. Many existing theories (e.g., probability theory, fuzzy numbers [30–33], Z-numbers [34–36], D-numbers [37–39], and Dempster–Shafer evidence theory) have been developed to represent various types of uncertainty. Dempster–Shafer evidence theory (D-S theory) can be regarded as a general extension of Bayesian theory. It was first proposed by Dempster in 1967 [40], and was developed to its present form by Glenn Shafer in 1976 [41]. D-S theory can present and handle uncertainty preferably than probability theory [42–44]. Moreover, it provides a combination rule to fuse different data from various information sources. D-S theory is now being studied for use in many fields, such as risk assessment [45,46], decision making [47–49], fault diagnosis [50,51], and others [52–56].

Let $\Theta = \{\{H_1\}, \{H_2\}, \cdots, \{H_n\}\}$ be a finite nonempty set of n elements that are mutually exclusive and exhaustive, 2^Θ is the power set composed of 2^n elements of Θ, and $\varnothing$ denotes the empty set. In D-S theory, mathematically a basic probability assignment (BPA) is a mapping: $2^\Theta \to [0, 1]$ that satisfies

$$\sum_{A \subseteq \Theta} m(A) = 1, m(\varnothing) = 0. \tag{1}$$

If m(A) > 0, A is called a focal element, and the set of all focal elements is named a body of evidence (BOE). The probability m(A) measures the belief exactly assigned to A and represents how strongly the evidence supports A. The mass of belief in an element of Θ is quite similar to a probability distribution, but differs in the fact that the unit mass is distributed among the elements of 2^Θ, not only on single subsets but on composite hypotheses as well. This is why BPA can better represent the uncertainty of information.

Moreover, a combination tool is offered by D-S theory to fuse multiple belief assignments as follows:

$$m = m_1 \oplus m_2 \oplus m_3 \oplus \cdots \oplus m_k, \tag{2}$$

where $\oplus$ represents the operator of combination. For two BOEs, Dempster's rule of combination is defined as:

$$m_1 \oplus m_2 = m(A) = \frac{\sum_{B \cap C = A} m_1(B)m_2(C)}{1 - K}, \tag{3}$$

where

$$K = \sum_{B \cap C = \varnothing} m_1(B)m_2(C). \tag{4}$$

The denominator, $1 - K$, is a normalization factor. K is called the degree of conflict between BOEs [57,58]. Dempster's rule strongly implies the agreement between diverse information and ignores the conflict between them. When information sources are in support of a similar proposition, it is able to reduce uncertainty in the combination result.

2.2. Analytic Hierarchy Process

Analytic hierarchy process (AHP), introduced by Thomas Satty (1980) [59], is an effective tool for dealing with complex decision making problem [60–63]. The weight determination process quantifies the subjective assessment of experts, and can check the consistency of decision-makers' evaluation. Generally, the process of applying AHP can be divided into three steps.

Step 1: Constructing the pair-wise comparison judgement matrix.

Assume that n pieces of decision elements are presented as $(F_1, F_2, F_3, \cdots, F_n)$. In order to compute the weight of decision elements, a comparison judgement matrix represented as $M_{n*n} = [m_{ij}]$ is created:

$$M_{n*n} = \begin{bmatrix} 1 & m_{12} & \cdots & m_{1n} \\ m_{21} & 1 & \cdots & m_{2n} \\ \vdots & \vdots & \ddots & \vdots \\ m_{n1} & m_{n2} & \cdots & 1 \end{bmatrix},$$

which satisfies:

$$m_{ij} = \frac{1}{m_{ji}}. \tag{5}$$

Each entry m_{ij} represents the importance of the ith element on the jth element. If $m_{ij} > 1$, then the ith element is more important than the jth element, while if $m_{ij} < 1$, then the ith element is less important than the jth element. If two elements have the same importance, then the entry m_{ij} is 1. The relative importance between two decision elements is measured according to a numerical scale from 1 to 9, as shown in Table 1.

Table 1. Pair-wise comparison scale for analytical hierarchy process (AHP) preference.

Value of m_{ij}	Interpretation
1	i and j are equally important
3	i is slightly more important than j
5	i is more important than j
7	i is strongly more important than j
9	i is absolutely more important than j
2, 4, 6, 8	intermediate values between the two adjacent judgements

Step 2: Calculating the weight of decision elements.

The eigenvector of M_{n*n} can be denoted as $\overline{\omega} = (\omega_1, \omega_2, \omega_3, \dots \omega_n)^T$, which is calculated using:

$$M\overline{\omega} = \lambda_{max}\overline{\omega}, \tag{6}$$

where λ_{max} is the largest eigenvalue of matrix M_{n*n}. The eigenvector corresponding to the largest eigenvalue can be viewed as the final criterion for ranking goals.

Step 3: Checking the consistency index.

A consistency index (CI) is used to measure the consistency within each pair-wise comparison judgement matrix, which is defined as:

$$CI = \frac{\lambda_{max} - n}{n - 1}. \tag{7}$$

Accordingly, the consistency ratio (CR) can be calculated as follows:

$$CR = \frac{CI}{RI}, \tag{8}$$

where RI is the random index. RI and CR are related to the dimension of the matrix, which is listed in Table 2.

Table 2. Values of the random index (RI).

Dimension	2	3	4	5	6	7	8	9	10
RI	0.00	0.58	0.90	1.12	1.24	1.32	1.41	1.45	1.51

Generally, if CR > 0.1, the consistency of the pair-wise comparison matrix M is unacceptable and the elements in the matrix should be revised. Otherwise, M is considered acceptable and the eigenvector ω is treated as the weighing vector after normalization.

There are three main advantages in AHP: simplicity, practicability, and systematicness. Simplicity refers to the fact that the computation to determine the weight of criteria is concise and the result is clear in order to make decision-making convenient. Practicability means that AHP can deal with a wide range of problems compared to traditional optimization methods with the combination of qualitative and quantitative analysis. Systematicness can be understood to describe how AHP comprehensively regards the object as a system and then makes decisions according to the decomposition, comparison, and judgement.

3. The Proposed Method

The Framework of the Proposed Method

The motivation behind the development of the proposed method is to propose a generalized method based on belief functions to evaluate the sustainable transport solutions under consideration. The proposed method addresses both objectives. One is that the flexibility is improved since the method can deal with uncertainty and conflictive information due to D-S theory. The other is that the modified method has good performance in sensitivity analysis. The framework of our proposed method is shown in Figure 1.

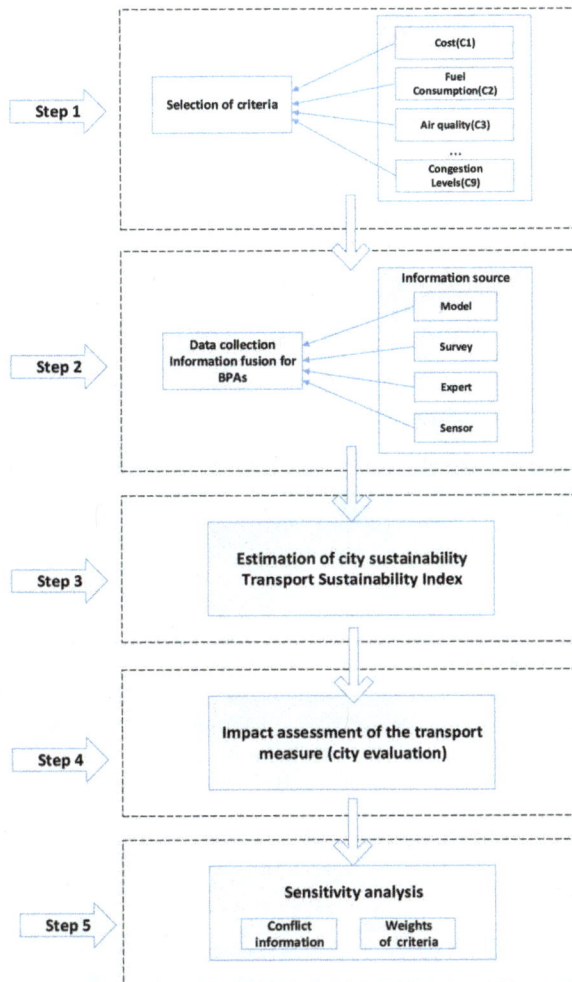

Figure 1. The framework of our proposed method. BPA: basic probability assignment.

Step 1: Selection of criteria.

The elements of sustainability refer to the Brundtland report of 1987 [64], and social, economic, and environmental indicators were concluded by transport researchers. The more detailed elements on three indicators are based on specific situations. In [20], nine evaluation criteria for sustainability assessment are introduced—namely, cost (C_1), fuel consumption (C_2), air quality (C_3), noise perception (C_4), users numbers (C_5), spatial accessibility (C_6), satisfaction (C_7), security (C_8), and congestion level (C_9), which are shown in Figure 2. AHP is used to determine the weight of different criteria, and detailed steps are represented in Section 2.2.

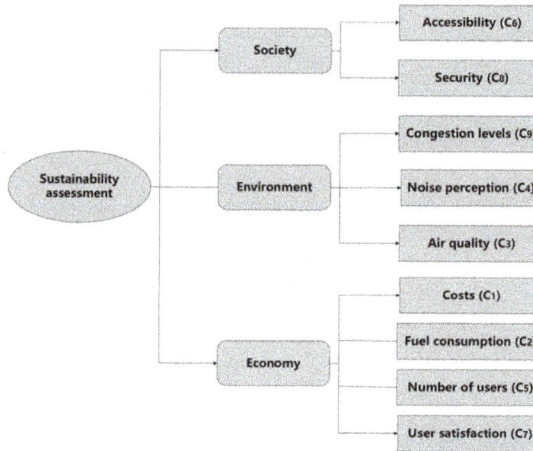

Figure 2. Evaluation criteria for sustainability assessment.

Step 2: Data collection and information fusion for BPAs.

After identifying evaluation criteria, data sets can be collected from various channels during the testing stage of the transportation measure. Anjali Awasthi et al. [20] mention four information sources: "Human experts", "Sensors", "Models" and "Survey". For example, experts usually have professional related experience or knowledge on city transportation to guarantee the reliability of information. Surveys are conducted with city residents and the responses are further aggregated to obtain BPAs. Models can be determined by the type of given data. For example, input data can be the number of votes that support a certain transport solution on different criteria, and thus we can apply a probability model to construct BPAs. Sensors use a measurement technique to allocate BPA values to different criteria for the transportation measure under study. With respect to each criteria, BPAs from different information sources are aggregated using Dempster's combination rule to generate a single belief function. The following will introduce the construction and representation of BPAs for collected data.

Experts allocate confidence to evaluation levels for each criteria. A new framework of evaluation level is proposed in this article: two evaluation levels, I and D. I means "increase" and D means "decrease". The vector of utility related to evaluation levels is given by u: u = {u(I), u(D), u(I,D)}. In other words, there exists a frame of discernment represented as $\Theta = \{I, D\}$, $2^{\Theta} = \{\{I\}, \{D\}, \{I, D\}\}$, the belief distribution on evaluation levels is represented with a form of mass function (i.e., BPA).

For example, in terms of the criteria "Cost" for car-sharing, a BPA from expert is shown as follows:

$$m\{I\} = 0.25; m\{D\} = 0.25; m\{I, D\} = 0.5.$$

This means that:

(1) "car-sharing solution will increase utility in terms of cost with the belief degree 0.25".
(2) "car-sharing solution will decrease utility in terms of cost with the belief degree 0.25".
(3) "the expert does not know whether the utility will increase or decrease with the belief degree 0.5".

Step 3: Utilities estimation.

For a certain transport solution, each criteria is either positively or negatively oriented with the utilities. For instance, a higher cost gets a lower utility while a higher air quality gets a higher utility. Complete utility values for nine criteria are shown in Table 3. Note that the number 1 represents the highest utility value, 0 represents the lowest utility value, and 0.3 represents an intermediate value chosen between 0 and 1. The global utility (u_i) for a criteria i can be calculated using the individual utility for evaluation levels $H_k \in \{I, D\}$ and the combined BPA. The formulation is as follows:

$$u_i = \sum_{k=1}^{p} u(H_k) \times BPA(H_k), \tag{9}$$

where p is equal to 3. $H_k = \{I, D, \{I,D\}\}$, $u(H_k)$ represents the individual utility of evaluation levels as shown in Table 3, and $BPA(H_k)$ is the basic probability assignment related to each evaluation level $u(H_k)$.

Table 3. Utility values for criteria [20].

Evaluation Criteria	Utility Values u(I), u(D), u(I,D)
Cost (C_1), Fuel consumption (C_2), Noise perception (C_4), Congestion level (C_9)	(0, 1, 0.3)
Air quality (C_3), Users numbers (C_5), Spatial accessibility (C_6), Satisfaction (C_7), Security (C_8)	(1, 0, 0.3)

Step 4: Estimation of city sustainability.

The global utilities for nine criteria can be used to estimate city sustainability at any given time t by a Transport Sustainability Index (TSI) [20]. Let us denote the global utilities for the criteria C_1, C_2, C_3, $\cdots$, C_N at time t_n by $u_1(t_n)$, $u_2(t_n)$, $u_3(t_n)$, $\cdots$, $u_n(t_n)$. As a result, the transport sustainability index (TSI) at time t_n is given by:

$$TSI(t_n) = u_1(t_n) \times \omega_1 + u_2(t_n) \times \omega_2 + u_3(t_n) \times \omega_3 + \cdots + u_n(t_n) \times \omega_n, \tag{10}$$

where ω_i is the weight of criteria obtained from AHP.

Step 5: Impact assessment of the transport measure.

The impact of the transportation measure on city sustainability is assessed by observing the change in TSI with respect to the pre-test and the post-test stages. Let t_{n-1} represents a time instant in the pre-test stage and t_n represent a time instant in the post-test stage, then the change in transport sustainability index over time interval $[t_{n-1}, t_n]$ is given by:

$$\triangle TSI[t_{n-1}, t_n] = TSI(t_n) - TSI(t_{n-1}), \tag{11}$$

If $\triangle$ TSI > 0, we can conclude that the impact of the transportation measure on city sustainability is positive and thus the measure can be regarded as sustainable transport. If $\triangle$ TSI < 0, the measure will be rejected for recommendation due to unsustainability.

Step 6: Sensitivity analysis.

To check the effectiveness of our proposed method, two experiments of conflictive information and changed weights of criteria will be illustrated in this part to better compare the difference between the method of Anjali Awasthi et al. [20] and ours.

The first experiment is about conflictive information to address the question "How sensitive is the overall decision to conflictive information during the information fusion process?" Conflictive information arises very easily due to man-made error, a fault in the model, or the subjectivity of experts' opinions. It is necessary to confirm that a method has redundancy to be more flexible and practical so that it has a more extensive application.

The second experiment is about the weight of criteria to address the question "How sensitive is the overall decision to small changes in the individual weight assigned during the pair-wise comparison process?" This question can be answered by slightly varying the values of weight and observing the effects on the decision. Moreover, because the weights of criteria are subjectively determined by experts using AHP, it is useful in situations where uncertainty exists in the definition of the importance of different factors.

4. Car-Sharing Application

Car-sharing means that a car is shared with many people. It is an alternative system of car ownership. In other words, the driver of vehicle only has the use right but no ownership. Car-sharing is similar to chartering a rental car for a short time. Clients reserve the vehicle in advance by telephone or online and then can get access to the vehicle. In general, a company is used to coordinating the vehicle and is responsible for the insurance and parking of vehicles. Car-sharing is a feasible option to reduce vehicle emissions by minimising the number of private vehicle movements inside cities [65–67]. The architecture of a typical car-sharing organization (CSO) is illustrated in Figure 3.

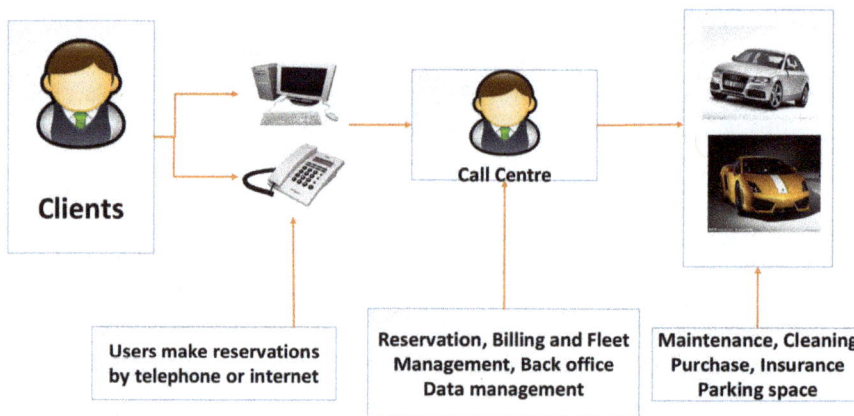

Figure 3. The architecture of a typical car-sharing organization (CSO) [20].

The aim of this part is to evaluate the influence of car-sharing on city sustainability and provide valuable suggestions for the city transportation authority to determine whether car-sharing can be implemented in the city.

4.1. Selection of Criteria

In the car-sharing system, nine criteria as mentioned previously are identified for sustainable evaluation: criteria cost (C_1), fuel consumption (C_2), air quality (C_3), noise perception (C_4), users numbers (C_5), spatial accessibility (C_6), satisfaction (C_7), security (C_8), and congestion level (C_9).

4.2. Data Collection and Information Fusion for BPAs

The data of BPAs on two evaluation levels (increase (I), decrease (D)) for nine criteria were collected from four information sources: expert, model, survey, and sensors/actual measurement. In order to provide convenience for calculation and comparison, note that the data at pre-test stage and post-test stage was from [20], yet there is a change in the framework of evaluation levels, and corresponding BPAs are shown respectively in Tables 4 and 5. In [20], three evaluation levels were introduced: I (increase), N (no change), D (decrease). In our article, only levels I and D are considered, and the original data corresponding to level N is mapped to the set {I,D}, in which BPA representation is formed.

Table 4. Data collection of BPAs at the pre-test stage [20].

Evaluation Criteria	Expert			Model			Survey			Sensors		
	I	D	Θ	I	D	Θ	I	D	Θ	I	D	Θ
cost (C_1)	0.25	0.25	0.5	0.6	0.2	0.2	0.5	0.2	0.3	0.5	0.15	0.35
fuel consumption (C_2)	0.3	0.3	0.4	0.4	0.2	0.4	0.4	0.2	0.4	0.2	0.5	0.3
air quality (C_3)	0.65	0.15	0.2	0.7	0.2	0.1	0.6	0.2	0.2	0.8	0.1	0.1
noise perception (C_4)	0.25	0.65	0.1	0.8	0.1	0.1	0.5	0.2	0.3	0.1	0.8	0.1
users numbers (C_5)	0.7	0.2	0.1	0.2	0.1	0.7	0.8	0.1	0.1	0.6	0.1	0.3
spatial accessibility (C_6)	0.5	0.3	0.2	0.6	0.1	0.3	0.5	0.2	0.3	0.7	0.1	0.2
satisfaction (C_7)	0.6	0.1	0.3	0.7	0.1	0.2	0.6	0.1	0.3	0.8	0.1	0.1
security (C_8)	0.4	0.3	0.3	0.4	0.2	0.4	0.4	0.2	0.4	0.5	0.3	0.2
congestion level (C_9)	0.4	0.4	0.2	0.1	0.5	0.4	0.2	0.5	0.3	0.2	0.6	0.2

Table 5. Data collection of BPAs at the post-test stage [20].

Evaluation Criteria	Expert			Model			Survey			Sensors		
	I	D	Θ	I	D	Θ	I	D	Θ	I	D	Θ
cost (C_1)	0.3	0.2	0.5	0.4	0.2	0.4	0.2	0.1	0.7	0.1	0.3	0.6
fuel consumption (C_2)	0.2	0.5	0.3	0.1	0.5	0.4	0.2	0.3	0.5	0.2	0.4	0.4
air quality (C_3)	0.6	0.1	0.3	0.7	0.1	0.2	0.6	0.2	0.2	0.7	0.2	0.1
noise perception (C_4)	0.1	0.6	0.3	0.1	0.8	0.1	0.2	0.7	0.1	0.2	0.6	0.2
users numbers (C_5)	0.8	0.1	0.1	0.7	0.1	0.2	0.6	0.3	0.1	0.8	0.1	0.1
spatial accessibility (C_6)	0.6	0.1	0.3	0.8	0.1	0.1	0.6	0.3	0.1	0.7	0.1	0.2
satisfaction (C_7)	0.7	0.1	0.2	0.8	0.1	0.1	0.6	0.2	0.2	0.6	0.1	0.3
security (C_8)	0.25	0.4	0.35	0.3	0.3	0.4	0.2	0.3	0.5	0.2	0.4	0.4
congestion level (C_9)	0.2	0.5	0.3	0.1	0.7	0.2	0.2	0.6	0.2	0.2	0.4	0.4

According to Equations (2)–(4), a comprehensive evaluation for each criterion can be calculated by Dempster's combination rule. Consider the criterion "Cost" in Table 5; let us denote the BPA from Expert by m_1^1, from Model by m_2^1, from Survey by m_3^1, and from Sensors by m_4^1. The detailed procedure for data combination can be shown as follows:

1. Original BPAs from Table 5 is:

$$m_1^1(I) = 0.3, m_1^1(D) = 0.2, m_1^1(\Theta) = 0.5,$$
$$m_2^1(I) = 0.4, m_2^1(D) = 0.2, m_2^1(\Theta) = 0.4,$$
$$m_3^1(I) = 0.2, m_3^1(D) = 0.1, m_3^1(\Theta) = 0.7,$$
$$m_4^1(I) = 0.1, m_4^1(D) = 0.3, m_4^1(\Theta) = 0.6.$$

2. Data information: Using Equations (2)–(4), we have $m_1^1 \oplus m_2^1 \oplus m_3^1 \oplus m_4^1 =$

$$m_1(I) = 0.5645,$$
$$m_1(D) = 0.3226,$$
$$m_1(\Theta) = 0.1129.$$

From the above analysis, a comprehensive BPA is obtained for the criteria "Cost" from four information sources. Likewise, the remaining BPAs for eight criteria can be computed. The aggregated BPA results for nine criteria at the pre-test and post-stage stages of car-sharing are represented in Table 6.

Table 6. The aggregated BPA results for nine criteria.

Evaluation Criteria	Pre-Test Stage			Post-Test Stage		
	I	D	Θ	I	D	Θ
cost (C_1)	0.8413	0.1365	0.0222	0.5645	0.3226	0.1129
fuel consumption (C_2)	0.5039	0.4488	0.0472	0.1234	0.8225	0.0541
air quality (C_3)	0.9831	0.0161	0.0008	0.9796	0.0189	0.0015
noise perception (C_4)	0.4260	0.5714	0.0026	0.0088	0.9908	0.0004
users numbers (C_5)	0.9680	0.0285	0.0035	0.9861	0.0132	0.0008
spatial accessibility (C_6)	0.9375	0.0550	0.0075	0.9907	0.0089	0.0004
satisfaction (C_7)	0.9855	0.0117	0.0027	0.9891	0.0098	0.0010
security (C_8)	0.7379	0.2388	0.0233	0.3566	0.5778	0.0657
congestion level (C_9)	0.1377	0.8503	0.0120	0.0343	0.9598	0.0060

4.3. Utilities of Estimation

After BPAs for nine criteria are obtained, global utility of each criteria on car-sharing can be calculated using individual utility values from Table 3. For example, the global utility for criteria "Cost(C_1)" and criteria "Air quality(C_3)" are computed using Equation (11):

$$u_1 = u(I) * BPA(I) + u(D) * BPA(D) + u(\Theta) * BPA(\Theta)$$
$$= 0 * 0.5645 + 1 * 0.3226 + 0.3 * 0.1129$$
$$= 0.3565$$

$$u_3 = u(I) * BPA(I) + u(D) * BPA(D) + u(\Theta) * BPA(\Theta)$$
$$= 1 * 0.9796 + 0 * 0.0189 + 0.3 * 0.0015$$
$$= 0.9800$$

Similarly, global utilities of the remaining seven criteria can be computed. The results are shown in Table 7.

Table 7. Global utilities for nine criteria.

Evaluation Criteria	At the Pre-Test Stage	At the Post-Test Stage
cost (C_1)	0.1432	0.3565
fuel consumption (C_2)	0.4630	0.8387
air quality (C_3)	0.9834	0.9800
noise perception (C_4)	0.5722	0.9909
users numbers (C_5)	0.9691	0.9863
spatial accessibility (C_6)	0.9398	0.9908
satisfaction (C_7)	0.9864	0.9894
security (C_8)	0.7449	0.3763
congestion level (C_9)	0.8539	0.9615

4.4. Estimation of City Sustainability

As mentioned before, a Transport Sustainability Index (TSI) can be used to measure city sustainability at any given time with global utilities. The Transport Sustainability Indexes of car-sharing at pre-test and post-test stages are denoted by TSI(t_{before}) and TSI(t_{after}), respectively. Assuming an equal weight of 0.111 for all criteria, using global utilities (Table 7) and criteria weight, we have:

$$TSI(t_{before}) = 0.1432 * 0.111 + 0.4630 * 0.111 + 0.9834 * 0.111$$
$$+0.5722 * 0.111 + 0.9691 * 0.111 + 0.9398 * 0.111$$
$$+0.9864 * 0.111 + 0.7449 * 0.111 + 0.8539 * 0.111$$
$$= 0.7388$$

$$TSI(t_{after}) = 0.3565 * 0.111 + 0.8387 * 0.111 + 0.9800 * 0.111$$
$$+0.9909 * 0.111 + 0.9863 * 0.111 + 0.9908 * 0.111$$
$$+0.9894 * 0.111 + 0.3763 * 0.111 + 0.9615 * 0.111$$
$$= 0.8292$$

4.5. Impact Assessment of the Transport Measure (Transport Solution Evaluation)

From the above, we can obtain the TSI values for car-sharing at pre-test and post-test stages and easily find that TSI(t_{after}) > TSI(t_{before}), 0.8292 > 0.7388. Therefore, a conclusion is drawn that the change brought by the transportation measure "car-sharing" is positive and thus car-sharing is suggested for adoption in the city.

4.6. Sensitivity Analysis

4.6.1. Experiment 1

From Tables 4 and 5, we can find that most BPAs have a similar trend to support the same evaluation level. For example, in Table 5 for the criteria "Air quality (C_3)", most BPAs from four information sources have the most confidence in supporting the evaluation level "I"; for the criteria "Congestion levels (C_9)", the collected data show that there is strong confidence with the evaluation level "D". However, in real life, it is noteworthy that incorrect data is possibly derived from human error, fault in the model, or the subjectivity of experts' opinion. Regardless, conflictive data may occur. Experiments were performed to determine the sensitivity of the final decision with respect to conflictive BPAs, and a comparison between Anjali Awasthi et al.'s method [20] and our proposed method is given.

From Table 5, for the criteria "Number of users", the model and sensor respectively provide a BPA: m(I) = 0.7, m(D) = 0.1, m(Θ) = 0.2; m(I) = 0.8, m(D) = 0.1, m(Θ) = 0.1. In this experiment, two new BPAs are substituted: m(I) = 0, m(D) = 0.1, m(Θ) = 0.9; m(I) = 0.9, m(D) = 0.1, m(Θ) = 0. The other two information sources "Expert" and "Survey" support the evaluation level "I" with a strong confidence $m_5^1 = 0.8$, $m_5^3 = 0.6$, while the new BPA from the model has the most confidence in supporting the set "Θ". Generally, it is called a piece of conflict evidence. Global utilities for nine criteria at the post-test stage with two methods are shown in Tables 8 and 9.

Table 8. The testing BPA fusion results.

Evaluation Criteria	Anjali Awasthi et al.'s Method			Our Proposed Method		
	I	D	Θ	I	D	Θ
cost (C_1)	0.0274	0.9589	0.0137	0.2258	0.5053	0.2689
fuel consumption (C_2)	0.0146	0.4380	0.5474	0.2296	0.3596	0.4108
air quality (C_3)	0.9910	0.0067	0.0022	0.5893	0.2087	0.2018
noise perception (C_4)	0.0020	0.0030	0.9951	0.2026	0.2399	0.5574
users numbers (C_5)	0	0	1.0000	0.6913	0.1894	0.1932
spatial accessibility (C_6)	0.9956	0.0030	0.0015	0.5946	0.2373	0.1681
satisfaction (C_7)	0.9931	0.0059	0.0010	0.5771	0.2566	0.1661
security (C_8)	0.0661	0.6167	0.3172	0.2529	0.3743	0.3727
congestion level (C_9)	0.0089	0.0536	0.9375	0.2290	0.3350	0.4359

Table 9. Global utilities of testing for nine criteria with two methods.

Evaluation Criteria	Anjali Awasthi et al.'s Method	Our Proposed Method
cost (C_1)	0.1432	0.3565
fuel consumption (C_2)	0.4630	0.8387
air quality (C_3)	0.9834	0.9800
noise perception (C_4)	0.5722	0.9909
users numbers (C_5)	0	0.9846
spatial accessibility (C_6)	0.9398	0.9908
satisfaction (C_7)	0.9864	0.9894
security (C_8)	0.7449	0.3763
congestion level (C_9)	0.8539	0.9615

From Table 10 and Figure 4, it can be easily seen that before modifying the data, a positive impact on urban transportation can be obtained by testing the transportation solution, while a negative impact is found after the emergence of conflictive evidence in Anjali Awasthi et al.'s method [20]. However, in our proposed method, we can come to the same conclusion that the influence brought by car-sharing in the city is positive and therefore car-sharing is recommended for adoption, which demonstrates that the proposed method can handle conflictive data well.

The Transport Sustainability Index (TSI) at the post-test stage for the two methods and the comparison of the two methods are shown in Table 10.

Table 10. Comparison of the two methods.

	Anjali Awasthi et al.'s Method	Our Proposed Method
TSI(t_{before})	0.7033	0.7388
TSI(t_{after})	0.6843	0.8190
△ TSI	TSI(t_{before}) > TSI(t_{after})	TSI(t_{before}) < TSI(t_{after})
Transport solution evaluation	Negative	Positive
Transport solution evaluation (without conflict)	Positive	Positive

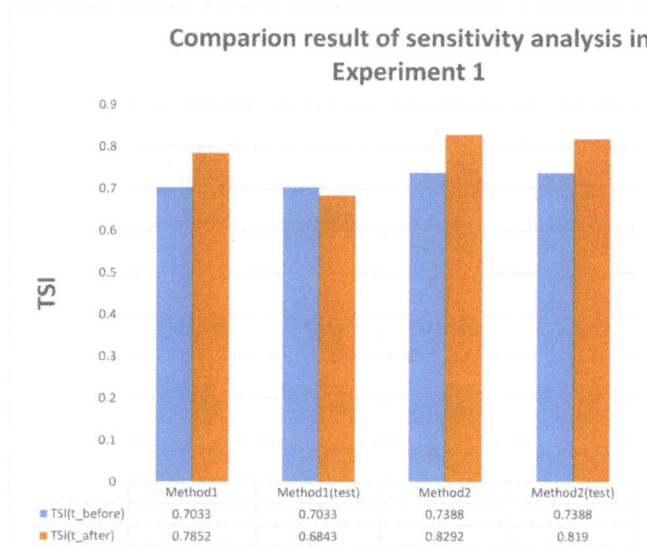

Comparion result of sensitivity analysis in Experiment 1

	Method1	Method1(test)	Method2	Method2(test)
TSI(t_before)	0.7033	0.7033	0.7388	0.7388
TSI(t_after)	0.7852	0.6843	0.8292	0.819

Figure 4. The first comparison in the sensitivity analysis of two methods. Method 1: Anjali Awasthi et al.'s method. Method 2: Our proposed method.

4.6.2. Experiment 2

In this article, equal weight for different criteria is assumed to calculate the TSI at pre-test and post-test stages for simplicity. Experiments were conducted to determine the sensitivity of the final decision with respect to weight changes of different criteria. These experiments are represented in Table 11.

Table 11. Different weight for nine criteria [20].

Experiment 2	Weights of Criteria								
	C_1	C_2	C_3	C_4	C_5	C_6	C_7	C_8	C_9
1	1	0	0	0	0	0	0	0	0
2	0	1	0	0	0	0	0	0	0
3	0	0	1	0	0	0	0	0	0
4	0	0	0	1	0	0	0	0	0
5	0	0	0	0	1	0	0	0	0
6	0	0	0	0	0	1	0	0	0
7	0	0	0	0	0	0	1	0	0
8	0	0	0	0	0	0	0	1	0
9	0	0	0	0	0	0	0	0	1
10	0	0	0.2	0	0.2	0.2	0.2	0.2	0
11	0.1	0.1	0.2	0.2	0.05	0.1	0.05	0.1	0.1
12	0.25	0.25	0	0.25	0	0	0	0	0.25
13	0.125	0.125	0.1	0.125	0.1	0.1	0.1	0.1	0.125
14	0.111	0.111	0.111	0.111	0.111	0.111	0.111	0.111	0.111
15	0.5	0.0625	0.0625	0.0625	0.0625	0.0625	0.0625	0.0625	0.0625
16	0.0625	0.5	0.0625	0.0625	0.0625	0.0625	0.0625	0.0625	0.0625
17	0.0625	0.0625	0.5	0.0625	0.0625	0.0625	0.0625	0.0625	0.0625
18	0.0625	0.0625	0.0625	0.5	0.0625	0.0625	0.0625	0.0625	0.0625
19	0.0625	0.0625	0.0625	0.0625	0.5	0.0625	0.0625	0.0625	0.0625
20	0.0625	0.0625	0.0625	0.0625	0.0625	0.5	0.0625	0.0625	0.0625
21	0.0625	0.0625	0.0625	0.0625	0.0625	0.0625	0.5	0.0625	0.0625
22	0.0625	0.0625	0.0625	0.0625	0.0625	0.0625	0.0625	0.5	0.0625
23	0.0625	0.0625	0.0625	0.0625	0.0625	0.0625	0.0625	0.0625	0.5

From Table 11, we can find that:

(1) Experiments 1–9 consider one criterion at a time with a maximum weight = 1 and allocate weight = 0 to the remaining eight criteria.

(2) Experiment 10 provides equal weight = 0.2 to the criteria with the high utility values for increase (I) for post-test stage (i.e., C_3, C_5, C_6, C_7, C_8). The weight of the remaining criteria is equal to 0.

(3) Experiment 11 gives a random allocation of weight to different criteria.

(4) Experiment 12 provides equal weight = 0.25 to the criteria with low utility values for increase (I) for the post-test stage (i.e., C_1, C_2, C_4, C_9).

(5) Experiment 13 distributes equal weight = 0.1 to criteria with high utility values for increase (I) for the post-test stage (i.e., C_3, C_5, C_6, C_7, C_8). Equal value = 0.125 is given to criteria with low utility values for increase (I) for post-test stage (i.e., C_1, C_2, C_4, C_9).

(6) Experiment 14 sets 0.111 as the weight of all criteria.

(7) Experiments 15–23 provide weight = 0.25 over one criteria and distribute the remaining 0.5 weight over eight criteria, making their criteria weight = 0.0625.

The results are presented in Figure 5.

From Figure 5, we can find that most experiments support the conclusion that the Transportation Sustainability Index (TSI) at the post-test stage is higher than the pre-test stage regardless of the changes in criteria weight, which shows the robustness of our proposed method. As a result, we can draw the conclusion that car-sharing has a positive impact on improving transport conditions inside the city. Car-sharing is recommended for adoption.

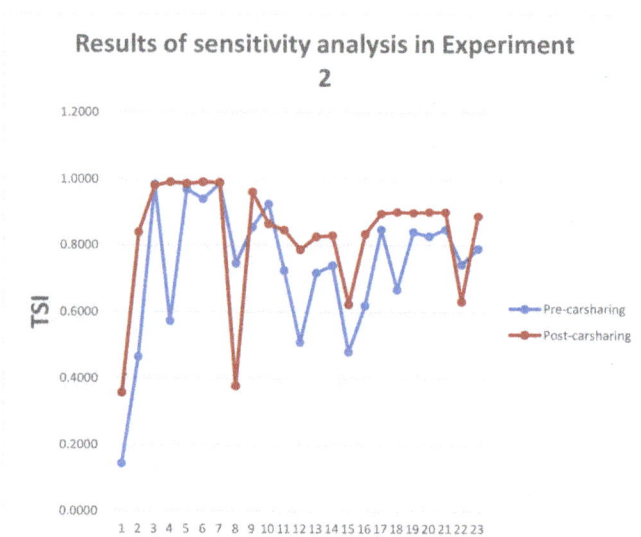

Figure 5. Results of sensitivity analysis in Experiment 2.

5. Discussion

In our proposed model, analytic hierarchy process (AHP) and Dempster–Shafer evidence theory (D-S theory) are combined to evaluate the impact of transport measures on city sustainability. The desirable properties of the model are stated as follows:

Basic probability assignment (BPA) rather than probability function is used to represent the confidence of evaluation levels for all criteria. In Anjali Awasthi et al.'s method [20], three evaluation levels {I, N, D} are mentioned which satisfies the equation p(I) + p(N) + p(D) = 1; while in our method, m(I) + m(D) + m(Θ) = 1, which means that the unit mass is distributed among the singletons in Θ and composite hypotheses as well. Through a modification, the probability function was transformed to BPA in D-S theory to better express the uncertainty. D-S theory has a strong function to manage uncertain information from various sources to make a decision. Furthermore, conflictive information can be handled well in our proposed method. As can be seen in Experiment 1, when conflictive data occur, two very different results are obtained in Anjali Awasthi et al.'s method [20], instead of the consistent result obtained in our proposed method.

Furthermore, in this article we come to the conclusion that car-sharing has a positive effect on city sustainability, which can be explained as follows: by reducing trips in private vehicle movements in the city, car-sharing can reduce vehicle emissions, as well as the occurrence of accidents and traffic congestion in the urban network to some extent. Moreover, car-sharing provides alternatives for humans' travel and thus can bring convenience. In a word, car-sharing can promote city sustainability through the improvement of urban conditions in terms of social and environmental aspects.

6. Conclusions

In this article, a modified method for evaluating sustainable transport solutions combining AHP and Dempster–Shafer evidence theory is proposed. AHP is adopted to determine the weight of sustainability criteria while D-S theory is used for data fusion of sustainability assessment. A Transport Sustainability Index (TSI) is presented as a primary measure to determine whether transport solutions have a positive impact on city sustainability. Finally, a case study of car-sharing is illustrated to show the efficiency of our proposed method. Compared with existing methods, two advantages can be listed: one is that the BPA is generated with a new modification framework of evaluation levels, which can flexibly address uncertain information. The other is that the modified method has excellent performance in sensitivity analysis. The model can be widely used in evaluating the impact of environment-friendly transport measures on city sustainability.

Acknowledgments: The authors greatly appreciate the reviewers' suggestions and the editor's encouragement. The work is partially supported by National Natural Science Foundation of China (Program Nos. 61703338, 61671384).

Author Contributions: Luyuan Chen and Xinyang Deng conceived and designed the experiments; Luyuan Chen performed the experiments; Luyuan Chen and Xinyang Deng analyzed the data; Luyuan Chen wrote the paper; and Xinyang Deng conducted the writing instructions.

Conflicts of Interest: The authors declare no conflict of interest.

References

1. Grant-Muller, S.; Usher, M. Intelligent Transport Systems: The Propensity for Environmental and Economic Benefits. *Technol. Forecast. Soc. Chang.* **2014**, *82*, 149–166.
2. Hsu, C.Y.; Yang, C.S.; Yu, L.C.; Lin, C.F.; Yao, H.H.; Chen, D.Y.; Lai, K.R.; Chang, P.C. Development of a Cloud-Based Service Framework for Energy Conservation in a Sustainable Intelligent Transportation System. *Int. J. Prod. Econ.* **2015**, *164*, 454–461.
3. Nita, M.R.; Badiu, D.L.; Onose, D.A.; Gavrilidis, A.A.; Gradinaru, S.R.; Nastase, I.I.; Lafortezza, R. Using Local Knowledge and Sustainable Transport to Promote a Greener City The Case of Bucharest, Romania. *Environ. Res.* **2018**, *160*, 331–338.
4. Chen, C.L. City-Hubs: Sustainable and Efficient Urban Transport Interchanges. *J. Transp. Geogr.* **2017**, *65*, 200–201.
5. Al-Atawi, A.M.; Kumar, R.; Saleh, W. Transportation Sustainability Index for Tabuk City in Saudi Arabia: An Analytic Hierarchy Process. *Transport* **2016**, *31*, 47–55.
6. Le Pira, M.; Inturri, G.; Ignaccolo, M.; Pluchino, A. Analysis of AHP Methods and the Pairwise Majority Rule (PMR) for Collective Preference Rankings of Sustainable Mobility Solutions. *Transp. Res. Procedia* **2015**, *10*, 777–787.
7. Gunarathna, P.; Hassan, R. Sustainability assessment tool for road transport asset management practice. *Road Transp. Res.* **2016**, *25*, 15–26.
8. Xu, S.; Jiang, W.; Deng, X.; Shou, Y. A modified Physarum-inspired model for the user equilibrium traffic assignment problem. *Appl. Math. Modell.* **2018**, *55*, 340–353.
9. Deakin M.; Curwell, S.; Lombardi, P. Sustainable urban development: the framework and directory of assessment methods. *J. Environ. Assess. Policy Manag.* **2002**, *4*, 171–197.
10. Parkin, J.; Wardman, M.; Page, M. Estimation of the determinants of bicycle mode share for the journey to work using census data. *Transportation* **2008**, *35*, 93–109.
11. Danilina, N.; Vlasov, D. Development of "Park-and-Ride" system as a tool for sustainable access control managing. In Proceedings of the Iop Conference Series-Earth and Environmental Science, Malang, Indonesia, 6–7 March 2017; Volume 90.
12. Shaukat, N.; Khan, B.; Ali, S.M.; Mehmood, C.A.; Khan, J.; Farid, U.; Majid, M.; Anwar, S.M.; Jawad, M.; Ullah, Z. A survey on electric vehicle transportation within smart grid system. *Renew. Sustain. Energy Rev.* **2018**, *81*, 1329–1349.
13. LucianoBarcellos de Paula, F.A.S. Algorithms applied in decision-making for sustainable transport. *J. Clean. Prod.* **2018**, *176*, 1133–1143.
14. Bovy, P.H.; Hoogendoorn-Lanser, S. Modelling route choice behaviour in multi-modal transport networks. *Transportation* **2005**, *32*, 341–368.
15. Sumalee, A.; Uchida, K.; Lam, W.H. Stochastic multi-modal transport network under demand uncertainties and adverse weather condition. *Transp. Res. Part C Emerg. Technol.* **2011**, *19*, 338–350.
16. Jourquin, B.; Limbourg, S. Equilibrium traffic assignment on large Virtual Networks: Implementation issues and limits for multi-modal freight transport. *Eur. J. Transp. Infrastruct. Res.* **2006**, *6*, 205–228.
17. Mihyeon Jeon, C.; Amekudzi, A. Addressing sustainability in transportation systems: definitions, indicators, and metrics. *J. Infrastruct. Syst.* **2005**, *11*, 31–50.
18. Litman, T.; Burwell, D. Issues in sustainable transportation. *Int. J. Glob. Environ. Issues* **2006**, *6*, 331–347.

19. Wellar, B. *Sampler of Commentaries on Methods and Techniques that Could Be Used in Making Decisions about Identifying, Adopting, or Implementing Sustainable Transport Practices*; Wellar Consulting Inc.: Ottawa, ON, Canada, 2009.

20. Awasthi, A.; Chauhan, S.S. Using AHP and Dempster–Shafer theory for evaluating sustainable transport solutions. *Environ. Modell. Softw.* **2011**, *26*, 787–796.

21. Aroian, L.A. The probability function of the product of two normally distributed variables. *Ann. Math. Stat.* **1947**, *18*, 265–271.

22. Deng, Y. Generalized evidence theory. *Appl. Intell.* **2015**, *43*, 530–543.

23. Xiao, F. An Improved Method for Combining Conflicting Evidences Based on the Similarity Measure and Belief Function Entropy. *Int. J. Fuzzy Syst.* **2018**, *20*, 1256–1266.

24. Xiao, F.; Zhan, C.; Lai, H.; Tao, L.; Qu, Z. New Parallel Processing Strategies in Complex Event Processing Systems with Data Streams. *Int. J. Distrib. Sens. Netw.* **2017**, *13*, 1–15.

25. Xu, H.; Deng, Y. Dependent Evidence Combination Based on Shearman Coefficient and Pearson Coefficient. *IEEE Access* **2018**, *6*, 11634–11640.

26. Deng, X.; Xiao, F.; Deng, Y. An improved distance-based total uncertainty measure in belief function theory. *Appl. Intell.* **2017**, *46*, 898–915.

27. Jiang, W.; Chang, Y.; Wang, S. A method to identify the incomplete framework of discernment in evidence theory. *Math. Probl. Eng.* **2017**, *2017*, doi:10.1155/2017/7635972.

28. Jiang, W.; Hu, W. An improved soft likelihood function for Dempster-Shafer belief structures. *Int. J. Intell. Syst.* **2018**, doi:10.1002/int.219809.

29. Jiang, W.; Wang, S. An Uncertainty Measure for Interval-valued Evidences. *Int. J. Comput. Commun. Control* **2017**, *12*, 631–644.

30. Deli Irfan, C.N. Intuitionistic fuzzy parameterized soft set theory and its decision making. *Appl. Soft Comput.* **2015**, *28*, 109–113.

31. Zheng, X.; Deng, Y. Dependence Assessment in Human Reliability Analysis Based on Evidence Credibility Decay Model and IOWA Operator. *Ann. Nuclear Energy* **2018**, *112*, 673–684.

32. Zhang, R.; Ashuri, B.; Deng, Y. A novel method for forecasting time series based on fuzzy logic and visibility graph. *Adv. Data Anal. Classif.* **2017**, *11*, 759–783.

33. Jiang, W.; Wei, B. Intuitionistic fuzzy evidential power aggregation operator and its application in multiple criteria decision-making. *Int. J. Syst. Sci.* **2018**, *49*, 582–594.

34. Aliev, R.; Alizadeh, A.; Huseynov, O.; Jabbarova, K. Z-Number-Based Linear Programming. *Int. J. Intell. Syst.* **2015**, *30*, 563–589.

35. Zadeh, L.A. A note on Z-numbers. *Inf. Sci.* **2011**, *181*, 2923–2932.

36. Kang, B.; Chhipi-Shrestha, G.; Deng, Y.; Hewage, K.; Sadiq, R. Stable strategies analysis based on the utility of Z-number in the evolutionary games. *Appl. Math. Comput.* **2018**, *324*, 202–217.

37. Bian, T.; Zheng, H.; Yin, L.; Deng, Y. Failure mode and effects analysis based on D numbers and TOPSIS. *Qual. Reliab. Eng. Int.* **2018**, doi:10.1002/qre.2268.

38. Xiao, F. An Intelligent Complex Event Processing with D Numbers under Fuzzy Environment. *Math. Probl. Eng.* **2016**, doi:10.1155/2016/3713518.

39. Deng, X.; Deng, Y. D-AHP method with different credibility of information. *Soft Comput.* **2018**, doi:10.1007/s00500-017-2993-9.

40. Dempster, A.P. Upper and lower probabilities induced by a multivalued mapping. *Ann. Math. Stat.* **1967**, *38*, 325–339.

41. Shafer, G. *A Mathematical Theory of Evidence*; Princeton University Press: Princeton, NJ, USA, 1976; Volume 1.

42. Yager, R.; Fedrizzi, M.; Kacprzyk, J. *Advances in the Dempster-Shafer Theory of Evidence*; Wiley: Hoboken, NJ, USA, 1994.

43. Zadeh, L.A. A Simple View of the Dempster-Shafer Theory of Evidence and Its Implication for the Rule of Combination. *AI Mag.* **1986**, *7*, 85.

44. Deng, X. Analyzing the monotonicity of belief interval based uncertainty measures in belief function theory. *Int. J. Intell. Syst.* **2018**, doi:10.1002/int.21999.

45. Neshat, A.; Pradhan, B. Risk assessment of groundwater pollution with a new methodological framework: Application of Dempster-Shafer theory and GIS. *Nat. Hazards* **2015**, *78*, 1565–1585.

46. Duan, Y.; Cai, Y.; Wang, Z.; Deng, X. A novel network security risk assessment approach by combining subjective and objective weights under uncertainty. *Appl. Sci.* **2018**, *8*, 428.

47. Deng, X.; Jiang, W. Dependence assessment in human reliability analysis using an evidential network approach extended by belief rules and uncertainty measures. *Ann. Nuclear Energy* **2018**, *117*, 183–193.

48. Liu, T.; Deng, Y.; Chan, F. Evidential supplier selection based on DEMATEL and game theory. *Int. J. Fuzzy Syst.* **2018**, *20*, 1321–1333.

49. He, Z.; Jiang, W. An evidential dynamical model to predict the interference effect of categorization on decision making. *Knowl.-Based Syst.* **2018**, doi:10.1016/j.knosys.2018.03.014.

50. Xiao, F. A Novel Evidence Theory and Fuzzy Preference Approach-Based Multi-Sensor Data Fusion Technique for Fault Diagnosis. *Sensors* **2017**, *17*, 2504.

51. Xiao, F.; Aritsugi, M.; Wang, Q.; Zhang, R. Efficient Processing of Multiple Nested Event Pattern Queries over Multi-Dimensional Event Streams based on a Triaxial Hierarchical Model. *Artif. Intell. Med.* **2016**, *72*, 56–71.

52. Zhang, Q.; Li, M.; Deng, Y. Measure the structure similarity of nodes in complex networks based on relative entropy. *Phys. A Stat. Mech. Appl.* **2018**, *491*, 749–763.

53. Yin, L.; Deng, Y. Measuring transferring similarity via local information. *Phys. A Stat. Mech. Appl.* **2018**, *498*, 102–115.

54. Deng, X.; Han, D.; Dezert, J.; Deng, Y.; Shyr, Y. Evidence combination from an evolutionary game theory perspective. *IEEE Trans. Cybern.* **2016**, *46*, 2070–2082.

55. Deng, W.; Lu, X.; Deng, Y. Evidential Model Validation under Epistemic Uncertainty. *Math. Probl. Eng.* **2018**, doi:10.1155/2018/6789635.

56. Liang, W.; He, J.; Wang, S.; Yang, L.; Chen, F. Improved cluster collaboration algorithm based on wolf pack behavior. *Clust. Comput.* **2018**, doi:10.1007/s10586-018-1891-y.

57. Liu, W. Analyzing the degree of conflict among belief functions. *Artif. Intell.* **2006**, *170*, 909–924.

58. George, T.; Pal, N.R. Quantification of conflict in Dempster-Shafer framework: A new approach. *Int. J. Gen. Syst.* **1996**, *24*, 407–423.

59. Saaty, T.L. *The Analytic Hierarchy Process: Planning, Priority Setting, Resources Allocation*; McGraw: New York, NY, USA, 1980.

60. Chin, K.S.; Wang, Y.M.; Poon, G.K.K.; Yang, J.B. Failure mode and effects analysis using a group-based evidential reasoning approach. *Comput. Oper. Res.* **2009**, *36*, 1768–1779.

61. Zhou, X.; Hu, Y.; Deng, Y.; Chan, F.T.S.; Ishizaka, A. A DEMATEL-Based Completion Method for Incomplete Pairwise Comparison Matrix in AHP. *Ann. Oper. Res.* **2018**, doi:10.1007/s10479-018-2769-3.

62. Jajac, N.; Knezic, S.; Babić, Z. Integration of multicriteria analysis into decision support concept for urban road infrastructure management. *Croat. Oper. Res. Rev.* **2010**, *1*, 74–82.

63. Deng, X.; Jiang, W. An evidential axiomatic design approach for decision making using the evaluation of belief structure satisfaction to uncertain target values. *Int. J. Intell. Syst.* **2018**, *33*, 15–32.

64. Brundtland, G.H. *Report of the World Commission on Environment and Development: "Our Common Future"*; United Nations: New York, NY, USA, 1987.

65. Awasthi, A.; Chauhan, S.S.; Hurteau, X.; Breuil, D. An analytical hierarchical process-based decision-making approach for selecting car-sharing stations in medium size agglomerations. *Int. J. Inf. Decis. Sci.* **2008**, *1*, 66–97.

66. Rohr, T.; Rovigo, M. Public service approach to car-sharing in mid-sized towns: The example of Belfort (France). *IET Intell. Transp. Syst.* **2017**, *11*, 403–410.

67. Fellows, N.; Pitfield, D. An economic and operational evaluation of urban car-sharing. *Transp. Res. Part D Transp. Environ.* **2000**, *5*, 1–10.

applied sciences

MDPI

Article

Dynamic Supply Chain Design and Operations Plan for Connected Smart Factories with Additive Manufacturing

Byung Do Chung [1],*, Sung Il Kim [2] and Jun Seop Lee [3]

[1] Department of Industrial Engineering, Yonsei University, 50 Yonsei-ro, Seodaemun-gu, Seoul 03722, Korea
[2] AI Platform Development Team, LG Uplus, 32 Hangang-daero, Yongsan-gu, Seoul 04389, Korea; sung1eo@hanmail.net
[3] Department of Industrial Engineering, Yonsei University, 50 Yonsei-ro, Seodaemun-gu, Seoul 03722, Korea; wnstjq322@naver.com
* Correspondence: bd.chung@yonsei.ac.kr; Tel.: +82-2-2123-3875; Fax: +82-2-364-7807

Received: 4 March 2018; Accepted: 5 April 2018; Published: 8 April 2018

Abstract: Interest in smart factories and smart supply chains has been increasing, and researchers have emphasized the importance and the effects of advanced technologies such as 3D printers, the Internet of Things, and cloud services. This paper considers an innovation in dynamic supply-chain design and operations: connected smart factories that share interchangeable processes through a cloud-based system for personalized production. In the system, customers are able to upload a product design file, an optimal supply chain design and operations plan are then determined based on the available resources in the network of smart factories. The concept of smart supply chains is discussed and six types of flexibilities are identified, namely: design flexibility, product flexibility, process flexibility, supply chain flexibility, collaboration flexibility, and strategic flexibility. Focusing on supply chain flexibility, a general planning framework and various optimization models for dynamic supply chain design and operations plan are proposed. Further, numerical experiments are conducted to analyze fixed, production, and transportation costs for various scenarios. The results demonstrate the extent of the dynamic supply chain design and operations problem, and the large variation in transportation cost.

Keywords: connected smart factories; additive manufacturing; dynamic supply chain design; flexibility; customized demand

1. Introduction

In the era of mass production, manufacturing companies have emphasized the efficiency of processes such as procurement, production, and logistics, in order to minimize the cost of over- and under-stocking. To improve productivity and reduce costs, companies have focused on economies of scale, and analyzed the tradeoffs between different processes. As the importance of personalized customer needs increases and the product life cycle became shorter, more companies are focusing upon customized products and flexibility. Production and supply chain management (SCM) systems, such as lean and agile systems, have been introduced by many companies [1,2].

The manufacturing system that provides customized products has been discussed and implemented. In particular, a manufacturing paradigm of mass customization has been in place since the late 1980s. Even though there has been controversy over the level of individualization [1], mass customization has been implemented by several important concepts, including product family architecture, reconfigurable manufacturing system, and delaying differentiation [3]. Now, we are entering a new era, which demands a new manufacturing paradigm, focusing on highly customized products, such as economies of one [4] and make-to-individual production [5]. Therefore, with the adoption of new information and communication technology (ICT), developing new production and SCM systems that support personalized manufacturing has become an matter of urgency [6].

Many governments and manufacturing companies have been developing more automated and flexible systems, in order to provide customized products. In Germany for example, the Industry 4.0 strategy was announced, and the concept of the smart factory was introduced by integrating new ICT technology and manufacturing systems. With advanced ICT technologies such as the Internet of Things (IoT), cloud computing, data analytics, and 3D printers, we are able to develop new manufacturing and supply chain systems that are tailored to meet individualized requirements [7]. In such environments, equipment in a smart factory is connected by the IoT, and monitored and shared in a cloud system. When multiple factories are connected and shared, they become connected smart factories, or a web of smart factories. In a traditional manufacturing system, it may be difficult to share processes and facilities; however, in an environment of connected smart factories, it is possible to share factories in real time and create new business models [8]. Customers and other players in a supply chain can be connected and efficiently communicate through a cloud-based system. To test and implement the new supply chain, the EU implemented the ManuCloud project, to share the manufacturing capacity of production networks in the cloud architecture [9]. In Korea, connected smart factories with 3D printers supporting personalized production have been built in Daejeon, Gwangju, and other cities, and are managed in the cloud system under the name "Factory as a Service" (FaaS) [10]. Figure 1a shows the layout of the smart factory. 3D printers and post-processing facilities are located at the side of the octagon, while a robot arm in the center moves products. The factory in each city has interchangeable processes which can be shared when connected to the cloud system. Figure 1b shows the FaaS cloud system with which customers can upload computer-aided design (CAD) files of products, collaborate with engineers, and request production. The motivation for this research is a desire to develop a dynamic supply chain design and planning applications for connected smart factories, with additive manufacturing.

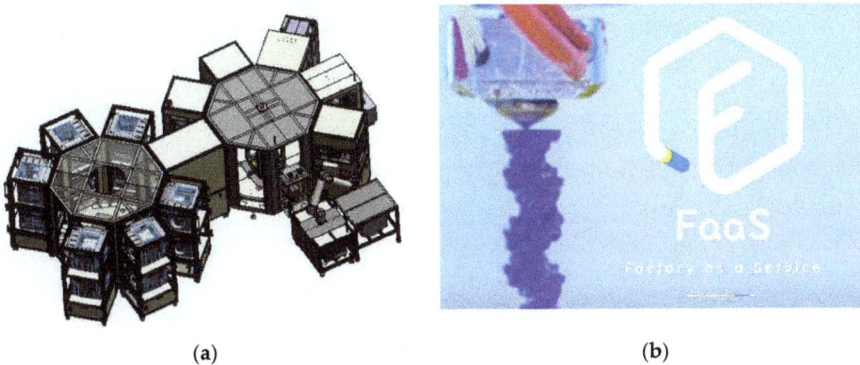

| (a) | (b) |

Figure 1. Connected smart factories and the cloud system. (**a**) Smart Factory; (**b**) Cloud System.

Theory and applications in SCM have been developed according to changes in manufacturing and business environments. Traditionally, the topics in SCM are classified into design and operation, and control aspects [11,12]. Supply chain design aspects deal with network configuration, outsourcing, and capacity decisions; these are long-term concerns for which changes in decisions have significant financial implications. Therefore, traditional supply chains are not frequently modified. Once a supply chain is designed according to the objective of the company, it is fixed for several months or years. Supply chain operation and control aspects are addressed after design decisions have been made. Medium and short term forecasting, production planning, and subcontracting are examples of operations and control issues. However, due to the aforementioned changes in business environments, a new approach is needed for dynamically designing supply chain, in order to efficiently support personalized production. In this paper, we consider a network design problem as a short-term decision based on real time data, including capacity and demand. The main contributions of the paper are as follows:

- In this paper, six types of flexibility associated with a network of smart factories utilizing 3D printers, cloud computing, and the IoT, are identified and defined. Specifically, design flexibility, product flexibility, process flexibility, supply chain flexibility, collaboration flexibility, and strategic flexibility are explained, based on a review of previous research.
- This paper proposes a general planning framework and two optimization models for supply chain design and operation, by dynamically connecting smart factories according to customer demand.
- This paper demonstrates a way of managing a network of smart factories to deal with customized products, and demonstrates the performance of the proposed approach with some scenarios.

The remainder of this paper is organized as follows. In Section 2, smart supply chains are described and the relevant literature is reviewed. Six types of flexibility are identified and explained with several examples. Papers considering mathematical models on supply chains with additive manufacturing are also reviewed. In Section 3, a general planning framework to solve the design and operations problems of a smart supply chain is presented. Optimization models are developed for selecting appropriate processes in the network of smart factories, and to generate efficient production and logistics plans using the selected processes. Numerical experiments are described in Section 4, and Section 5 concludes the paper.

2. Smart Supply Chain with Additive Manufacturing

2.1. The Concept of Smart Supply Chains

As customized products increase, companies require the ability to produce a variety of products and form an integrated network in order to efficiently utilize geographically distributed resources. A network providing customized products according to customer needs was proposed as the smart supply chain concept [13]. Later, Noori and Lee [14] focused on the role of small- and medium-sized enterprises (SMEs) in competing with big companies. Geographically spread but electronically linked SMEs could form a dynamic and adaptive network without obligatory egalitarian responsibility, which is different from traditional collaborative networks. The dynamic and adaptive network is referred to as a dispersed manufacturing network. Similar concepts of connecting companies to generate collaborative and competitive organizations include collaborative networks [15,16], responsive supply chains [17], and distributed manufacturing [18].

The concept of the smart supply chain is not new and has been discussed by many researchers and practitioners; however, recent research emphasizes the role of advanced ICT. Gaynor et al. [19] is one of the earlier works that proposes a smart supply chain using wireless sensors. As an example, they developed a prototype application to handle Sears' customized orders. Bendavid and Cassivi [20] suggest that the next step in the development of wireless sensors is the higher-level integration of inter-organizations, and e-commerce within a self-managed smart supply chain. In subsequent research, Bendavid and Cassivi [21] show how radio-frequency identification (RFID) could be adopted to implement smart supply chain models to deal with their dynamic nature. Ivanov and Sokolov [22] consider the next generation supply chain as a cyber-physical system (CPS)—an advanced network using a physical system and cloud service. They emphasize that a pre-determined supply chain structure will evolve into a dynamic and temporary network. The customer-centered new supply chain is more flexible, adaptable, and intelligent, so that it could operate without human involvement.

Butner [23] claims that a smarter supply chain has three properties: (1) instrumented—more data in a supply chain would be generated by various devices such as sensors, RFID, and actuators;

(2) interconnected—more objects in a supply chain would be extensively connected and lead to massive collation among them; and (3) intelligent—more intelligent systems would help people by making real-time decisions and predicting future events. Recently, Wu et al. [24] added three additional characteristics, including automated, integrated, and innovative.

This paper focuses on the flexibility and dynamics required to make customized products, which could be achieved by employing new ICT such as sensors, 3D-printers, CPS, and IoT. It considers a set of geographically dispersed smart factories sharing their machines and collaborating to make customized products through a cloud system—called connected smart factories. They are able to share machines through entire processes, or partially share some processes to make a customized product. Then, smart supply chains could be dynamically organized to fulfill a customer order and dissolve after order fulfillment, using the advanced ICT mentioned above.

2.2. Flexibility in the Smart Supply Chain with Additive Manufacturing

With advanced ICT, the smart supply chain could be more flexible than the traditional supply chain. There are many ways to implement a smart supply chain. For example, traditional production facilities with smart software systems can be a part of a smart supply chain. With anticipatory shipping models, a company is able to deal with individual customer's requirement. In this paper, we focus on the flexibilities achieved by connected smart factories with additive manufacturing that can be shared via a cloud system. A literature review was conducted to identify the flexibilities, which have been classified into six categories. Note that the purpose of the paper is not to thoroughly review the smart supply chain literature, but to identify smart supply chain flexibilities with 3D printing and other interchangeable processes.

2.2.1. Design Flexibility

A 3D printer using additive manufacturing technology adds layers using raw materials such as polymers, ceramics, and metals [25]. 3D printers can produce products in the desired shape without molds when the raw materials necessary for the production are ready; the variety of product shapes is therefore very broad. Design flexibility refers to the flexibility to produce a product with a variety of designs. 3D printers can be used to produce customized products with various designs performing the same function [4].

Production through 3D printouts can be based on CAD drawings, and do not require expensive tools used for subtractive manufacturing processes such as drilling, grinding, and molding [26]. The process of product design is less constrained by manufacturing processes, and has a high degree of freedom. With 3D printers, hollow-core structures are easily formed while maintaining or optimizing the characteristics and performance of products. For example, Lu et al. [27] propose a honeycomb-cell structure providing strength in tension with minimal material cost. Qin et al. [28] investigate the properties of spider web design. These new types of design allow companies to make products that perform the same function while reducing the weight of the parts compared to solid type designs, through topology optimization [29,30]. Moreover, in order to manufacture lighter parts, fewer raw materials are used and a manufacturer can save on raw materials costs. Lighter products consume less energy and produce less CO_2, and are therefore of significant interest to companies in automobiles and aviation.

2.2.2. Product Flexibility

Product flexibility means that a variety of products can be produced in one factory or supply chain according to customers' requirements. This concept is closely related to mass customization [1], but with some differences. While a traditional mass customization system is based on modular design or product postponement [31], mass customization in the smart supply chain is based on CPS and the Internet of Things and Services in the manufacturing system [7]. In addition, in contrast to traditional systems, it does not require a high degree of supply chain integration [26].

3D printers also play an important role in meeting individual customer requirements. In the past, 3D printers were mainly used for prototype and mockup production at low cost. However, due to the development and spread of technology, 3D printers are increasingly being used. Rather than using a single product, companies can test market response with small quantities of finished products of various sizes, colors, and functions [7,32]. In addition, 3D printers are successfully used to produce customized final products with individual product designs without retooling [29,33]. In this way, it is of the greatest advantage to be able to flexibly produce small quantities of various products without incidental tooling costs through 3D printing [34,35]. If the raw materials of the products to be printed are the same, any product could be produced in a factory.

2.2.3. Process Flexibility

Flexibility is needed to adapt to the wide variety of changes that occur at the process level of the production site in the manufacture of personalized products [36]. Process flexibility means that the same product can be produced through different manufacturing processes.

With the production line using 3D printers, process flexibility can be achieved so that one product can be produced in various ways. In other words, depending on the equipment or resources that are available, manufacturers could flexibly select the best process to make the product. First, the product can be redesigned to produce finished products with fewer components, requiring less assembly processes [37,38]. For example, GE Aviation used to produce fuel nozzles through the assembly of 20 parts. Now, they produce them as single units using additive manufacturing. Secondly, the performance of 3D printers could also affect the selection of the manufacturing process. For example, a large 3D printer produces a single product at a time, while a small 3D printer produces parts of the product that are assembled in the subsequent process. Thirdly, the manufacturing process depends on the selection of material. Typically, polymers require a limited finishing process while metals require post-processes [4].

2.2.4. Supply Chain Flexibility

The development of advanced technology can change the supply chain configuration, or could require the adoption of a new supply chain model [26,35]. Supply chain flexibility means that facilities in the connected smart factories are shared and operated flexibly. In the smart factory environment, supply chain flexibility can be achieved through outsourcing of manufacturing, which has become easier with the use of 3D printers [26], real time data collection, and agile collaboration between intelligent agents [39]. In addition, cyber-supported collaboration infrastructure plays an important role in sharing demand and capacity among enterprises in a collaborative network [40].

Further, the large variety of products that result from customization introduces significant uncertainty. It is possible to flexibly cope with such high supply chain uncertainties by sharing the capacity of the facilities in the network [41]. With advanced technology, collaborative manufacturing is considered to be a new business model [42]. If the capacity of a particular process in the factory is insufficient, a product can be made by using the capacity of other factories in the network, through sharing processes and machines. That is, available resources in the supply chain network can be used efficiently, and the capacity utilization of factories in the network can be maximized [43].

2.2.5. Collaboration Flexibility

Previous studies have emphasized that collaboration through information sharing within the supply chain has a significant impact on supply chain performance [44,45]. In recent years, with the introduction of new technologies and extensive connectivity, the amount and speed of information generated in the supply chain have become more diverse and faster than ever before. Product design, production, and product tracking information is collected in various ways, including RFID and the IoT, and is shared in a cloud. The cloud system monitors and stores information generated within the supply chain to provide visibility, and enables end-to-end collaboration through information sharing [46]. Collaboration flexibility means that supply chain participants can exchange feedback with each other and work together to provide customized products. It focuses on the integration of the design, process planning, production planning, and logistics, through information sharing, while supply chain flexibility focuses on the integration of supply stages through the sharing of available capacities.

In the traditional manufacturing environment, the roles and responsibilities of the designer and manufacturer are clearly distinguished, but in the new environment, the distinction becomes unclear [4]. In a smart supply chain, participants can collaborate from design to delivery; it extends the range of collaboration compared to the traditional approach, such as the use of concurrent engineering in the design stage. For example, even during production, a designer can request manufacturing process changes and a manufacturer can request product design changes according to the availability of equipment. Even customers who do not know how to draw a CAD model or produce a product can request production based on a purchased CAD file. Therefore, collaboration among supply chain participants is critically important; a smart supply chain should be able to support last-minute changes to individual customer requirements [7].

2.2.6. Strategic Flexibility

Strategic flexibility refers to the ability to respond in a timely manner to changes in market competition, and to environmental changes in an appropriate manner [47]. Many research studies have been conducted on strategic flexibility. Earlier works on strategic flexibility emphasized responsiveness. They focused on identifying characteristics and acquiring appropriate resources according to the changing environment. Appropriate implementation options were then developed [48,49]. In addition, enterprises are required to develop strategic diversity and then choose an appropriate strategy to best suit the changing environment [50,51].

When it comes to the concept of the smart supply chain with additive manufacturing, new products can be quickly introduced in response to market demand [52]. Also, timely response may be achieved through coordination of flexible resources connected to cloud-based networks in response to internal or external environmental changes. Shirodkar and Kempf [53] proposed a method to adopt flexible strategies to solve the problems of internal or external uncertainties in semiconductor production. They shared the capacity of each process facility to make the plant more flexible. Generally, one product is produced in one factory, but if there is insufficient capacity due to various uncertainties, it can be produced in another factory registered in the network. Seok and Nof [43] made a supply chain network operate flexibly by maximizing the capacity utilization of factories in the network.

2.3. Mathematical Models for Supply Chains with Additive Manufacturing

This section summarizes the relevant research on quantitative approaches to supply chains with additive manufacturing. Recent research focuses on the adoption of additive manufacturing and its effects. Scott and Harrison [52] consider additive manufacturing in an end-to-end supply chain, to compare the performance of additive and traditional manufacturing. They propose a stochastic optimization model and show that demand is the most important factor for the adoption of new technology. Barz et al. [54] investigate the impact of additive manufacturing on the structure of the supply chain. A two-stage capacitated facility location problem is used for supply chain configuration, consisting of suppliers, manufacturers, and customers. They show that as the efficiency of the new technology increases, the effect on the supply chain configuration is significant in three scenarios: evenly distributed facilities, clustered suppliers, and clustered manufacturers. Emelogu et al. [55] propose a two-stage stochastic programming model to analyze the cost, and to investigate the economic feasibility of additive manufacturing facilities. There are few studies on the operational aspects of the smart supply chain. Ivanov et al. [56] consider a mathematical model and algorithm for a scheduling problem of a smart factory, under the consideration of different processing speeds and dynamic job arrivals.

3. Materials and Methods

3.1. Planning Framework

Different to the mass production system, it may be more difficult or impossible to forecast highly customized products as customers and products are not known before orders are placed. Players in a supply chain providing customized products are able to know the product characteristics and the required manufacturing processes after receiving the order; these differ by product. Therefore, flexibility is more important than ever before.

Figure 2 shows the process for the smart supply chain from customer order to delivery of the product. First, the customer uploads a CAD file and requests an order via the cloud system using a website. Then, engineers supporting the system validate the CAD file, confirm whether the product can be made in the system, and provide feedback to the customer. At the same time, the required processes and bill of materials (BOM) are designed. Next, based on the design data, real time data about the machines in the processes are gathered from the IoT and monitored through the cloud system. After the required information—including availability of raw materials, capacity of resources, and working calendars—has been reviewed, the optimal supply chain design and operations plans are generated for a single order request by matching demand requirements and supply capability. Based on the decision, customers receive information on the facilities and processes that will be used, the production plan, and the delivery date. After confirmation from the customer, an operations plan is sent to the selected parties. Finally, each party in the supply chain develops his or her own schedule, and execution begins.

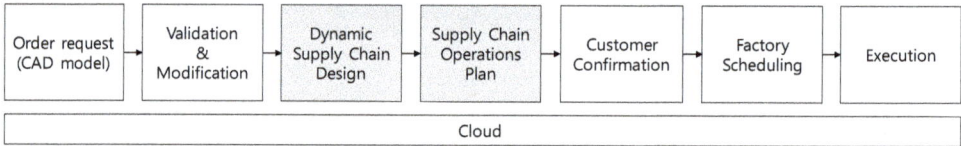

Figure 2. Smart supply chain planning process.

The optimization models in the following sections covers the steps for the development of the design and operations plan of the supply chain, which can be dynamically organized and deleted according to customer orders. It is assumed that the factories, as well as processes, are interchangeable. The example in Figure 3 explains the approach for the dynamic supply chain design and operation with connected smart factories, as proposed in this paper.

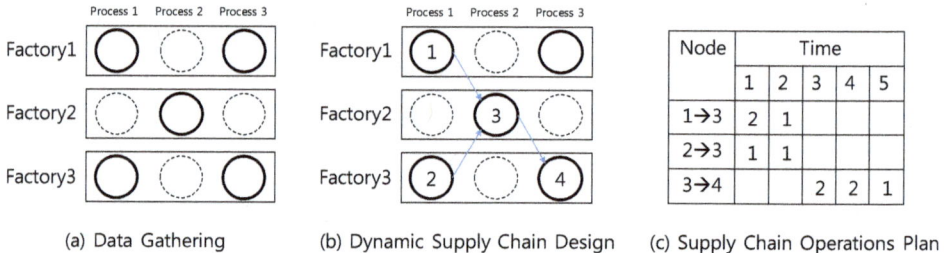

Figure 3. Dynamic supply chain design and operation.

In the first step, based on the data gathered from the customer and factories, all possible processes in each factory are selected, as outlined in Figure 3a. In the example, three processes, namely Processes 1, 2, and 3, are required to produce the product. In addition, there are three available factories, namely

Factories 1, 2, and 3, in the network. The nodes, represented by circles, show which process in which factory is available. A solid line indicates that the process in the factory is available, while a dotted line indicates unavailability. For example, Processes 1 and 3 in Factory 1 are available. After checking the availability, the optimal supply chain is designed. That is, the best combination of processes is selected in order to minimize the total cost and cover all required processes as shown in Figure 3b. Four processes in three factories are selected. That is, all three factories are sharing their resources to efficiently fulfill the current order. Process 3 in Factory 1 is not required to make the product, even if it is available. In the second step, optimal production and delivery plans are calculated using the processes selected in the previous step. Figure 3c shows the plan for each process.

3.2. Dynamic Supply Chain Design

The purpose of the dynamic supply chain design phase is to select the optimal combination of nodes to be used. We developed an integer programming model by modifying a set-covering problem. The concept is that all the required processes must be covered with at the minimum total cost. We assume that there are enough raw materials and no work-in-progress (WIP) in the process. The notations used in the optimization model are summarized in Table 1.

$$min \sum_{k \in K} FC_k x_k + \sum_{i \in I} \sum_{k \in K} \left(SC_{ik} y_{ik} + PC_{ij} q_{ik} \right) + \sum_{i \in I} \sum_{k \in K} \sum_{l \in K} \frac{PT_{ik} q_{ik}}{CAP_{ik}} TC_{jk} v_{ikl} \tag{1}$$

$$\text{s.t.} \qquad \sum_{k \in K} q_{ik} \geq d \quad \forall i \tag{2}$$

$$\sum_{k \in K} a_{ik} x_k \geq 1 \qquad \forall i \tag{3}$$

$$y_{ik} \leq a_{ik} x_k \qquad \forall i, k \tag{4}$$

$$p_{ik} q_{ik} \leq u_{ik} CAP_{ik} y_{ik} \qquad \forall i, k \tag{5}$$

$$v_{ikl} \leq y_{ik} \qquad \forall i, k, l \tag{6}$$

$$v_{ikl} \leq y_{i+1l} \qquad \forall i, k, l \tag{7}$$

$$v_{ikl} \geq y_{il} + y_{i+1l} - 1 \qquad \forall i, k, l \tag{8}$$

$$x_k, \ y_{ik} \in \{0, 1\} \quad \forall i, k \tag{9}$$

$$q_{ik} \geq 0, \ v_{ikl} \geq 0 \quad \forall i, k, l \tag{10}$$

Table 1. Notations for dynamic supply chain design.

Sets	
I	Set of processes or nodes ($i \in$ I)
K	Set of factories ($k, l \in$ K)
Parameters	
D	Demand for the product
CAP_{ik}	Average time available per day for process i in factory k in the planning horizon
u_{ik}	Utilization of process i in factory k
FC_k	Fixed cost for selecting factory k
SC_{ik}	Setup cost for process i in factory k
PC_{ik}	Process cost per unit for process i in factory k
PT_{ik}	Processing time per unit for process i in factory k
TC_{kl}	Transportation cost from factory k to factory l
a_{ik}	Set-covering matrix representing the relationship between process i and factory k
Decision Variables	
x_k	Factory selection, 1 if factory k is selected; 0, otherwise
y_{ik}	Process selection; 1 if process i in factory k is selected; 0, otherwise
v_{ikl}	Transportation selection; 1 if product from process i in factory k is sent to factory l
q_{ik}	Production quantity of process i in factory k

The objective function is expressed by Equation (1): it minimizes the total costs composed of fixed, production, and transportation costs. Fixed costs are incurred when a factory is selected, and the production costs are expressed as the sum of the setup cost and the production cost per unit multiplied by the production quantity. Finally, transportation costs arise when consecutive processes are carried out in different factories. Equation (2) represents the demand for a product, which requires a supply chain design. Equation (3) requires that all processes be covered, and Equation (4) indicates that the available process at a factory can be selected only if the factory is selected. Equation (5) represents a capacity constraint, and capacity is calculated in terms of available time. Equations (6) and (8) determine whether transportation between the factories is necessary or not. We note that the decision variable v_{ikl} is not necessarily defined as a binary variable within the constraints. Equations (9) and (10) define the decision variables.

3.3. Dynamic Supply Chain Operation

Once the supply chain has been designed, the operations plan is generated. Nodes represent processes selected in the dynamic supply chain design phase. We also introduce a dummy node, I_d, so that finished products manufactured in several factories are delivered to the customer. The supply chain operations model, using the notations in Table 2, is as follows.

$$min \sum_{i \in I} \sum_{j=I_d} \sum_{t \in T} p^t a_{ij} x_{ij}^t \tag{11}$$

$$\text{s.t.} \qquad y_i^t \geq c_i^t \qquad \forall\, i,t \tag{12}$$

$$II_i^t = II_i^{t-1} + \sum_{j \in I} a_{ji} x_{ji}^{t-LT_{ji}} - y_i^t \quad \forall\, i,t \tag{13}$$

$$IO_i^t = IO_i^{t-1} + y_i^t + \sum_{j \in I} a_{ij} x_{ij}^t \quad \forall\, i,t \tag{14}$$

$$\sum_{i \in I} \sum_{t \in T} a_{iI_d} x_{iI_d}^t \geq d \tag{15}$$

$$x_{ij}^t = 0 \quad \forall\, (i,j) \in V, t \in T \backslash T_d \tag{16}$$

$$x_{ij}^t, y_i^t, II_i^t, IO_i^t \geq 0 \quad \forall\, i,j,t \tag{17}$$

Table 2. Notations for dynamic supply chain operation.

Sets	
I	Set of processes or nodes ($i, j \in$ I)
T	Time periods (t $\in$ T)
T_d	Time periods when transportation model is available. ($T_d \subset$ T)
V	A pair of nodes connecting two consecutive processes in different factories
Parameters	
a_{ij}	Adjacency matrix resenting the relationship between processes
p^t	Penalty cost having an incremental function for time
c_i^t	Available capacity of node i at time t
LT_{ij}	Transportation lead time from node i to node j
d	Demand for the product
Decision Variables	
x_{ij}^t	Quantity of product sent from node i to node j at time t
y_i^t	Quantity of product processed in node i at time t
II_i^t	Raw material inventory level of node i at time t
IO_i^t	Manufactured product inventory level of node i at time t

The objective function, Equation (11), aims to ensure that the production is finished as soon as possible by employing penalty cost, which increases with time. The objective function can be modified according to the decision maker's goal. Equation (12) limits the production quantity depending on the capacity of the process at each time unit. Equations (13) and (14) represent inventory balance equations for raw material and manufactured product at each process, respectively. LT_{ij} represents the transportation lead time between processes i and j. Equation (15) is a constraint expressing the demand of the product. Equation (16) shows that inter-factory movement is only possible when the transportation mode is available. The manufactured product in a factory can be sent to other factories when t belongs to T_d. For example, all factories operate ten hours a day, and inter-factory transportation takes place at night after work hours. Equation (17) defines the decision variables.

4. Numerical Experiments

4.1. Data

Parameters related to factories and processes are randomly generated, and the data used in the numerical experiments are shown in Table 3. The processing time of the 3D printers is relatively longer than that of other processes such as painting, fumigating, and vision testing. It is assumed that the planning horizon is five days, and factories are operated 10 hours per day. Further, transportation between factories occurs at the conclusion of the 10 h operating day. All experiments were conducted using GAMS; the first mathematical model was solved using Baron, and the second by using CPLEX. The experiments used an Intel Core i5-6600 3.3 GHz processor and a computer with 8 GB memory.

Table 3. Data used in the experiments.

CAP_{ik}	[144,000, 180,000]	FC_k	[5000, 10,000]
u_{ik}	[0.8, 1]	SC_{ik}	[1000, 2000]
PT_{ik} (3D printing)	[2000, 4000]	PC_{ik}	[30, 50]
PT_{ik} (Others)	[500, 1000]	TC_{kl}	[2000, 5000]

4.2. Experimental Results

First, six connected factories and a product requiring four processes were considered. Table 4 shows a set-covering matrix for a small-scale problem. For example, at the time of demand occurrence, the available processes in connected smart factory 1 are 1,2, and 3.

Table 4. Set covering matrix.

a_{ik}		Factory					
		1	2	3	4	5	6
process	1	1	1	0	0	1	1
	2	1	0	1	0	1	0
	3	1	0	1	0	0	1
	4	0	1	1	1	0	0

Figure 4 shows the results of dynamic supply chain design. A comparison of the demand of 1 for an individual request and the demand of 100 for the marketing test shows that different supply chains are obtained even if all data, except demand, are the same. In the case of an individual request, Factory 1 is selected for Process 1 and Factory 3 for the remaining processes. However, in the case of the marketing test, Factories 1, 2, and 3 are selected and Processes 1 and 3 are performed in parallel at multiple factories. After completing the supply chain design, a production and transportation plan is obtained with the dynamic supply chain operation model. In the first case, Process 1 is performed on the first day and the remaining processes are performed on the second day, as shown in Table 5.

The customer can receive the product after two days. In the second case, demand is fulfilled after seven days, as shown in Table 6.

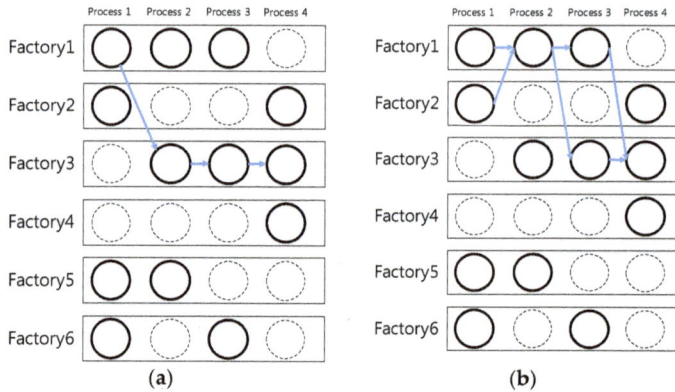

Figure 4. Dynamic supply chain design and operation. (a) Individual request; (b) Demand for marketing test.

Table 5. Production plan–Individual request.

	Day 1	Day 2	Day 3
Process 1 (Factory 1)	1	0	0
Process 2 (Factory 2)	0	1	0
Process 3 (Factory 2)	0	1	0
Process 4 (Factory 2)	0	1	0

Table 6. Production plan–Demand for marketing test.

	Day 1	Day 2	Day 3	Day 4	Day 5	Day 6	Day 7
Process 1 (Factory 1)	10	10	10	10	10	3	0
Process 1 (Factory 2)	10	10	10	10	7	0	0
Process 2 (Factory 1)	9	20	20	20	19	12	0
Process 3 (Factory 1)	7	9	10	10	10	10	0
Process 3 (Factory 3)	0	2	8	10	10	10	4
Process 4 (Factory 4)	0	9	16	20	20	20	15

4.3. Comparison of Results

In this section, the results are compared, considering problem size and capacity change. First, the problem size is increased by introducing more factories and more processes into the process whereby production takes place. The number of connected smart factories is increased to 12, in two separate units. In addition, the number of processes is increased to 10, in two units. Demand is fixed at 100. In all cases, the computation time required to solve each problem was less than one second. In Figure 5, the number before the parentheses represents the number of connected smart factories, and the number within parentheses represents the number of processes required.

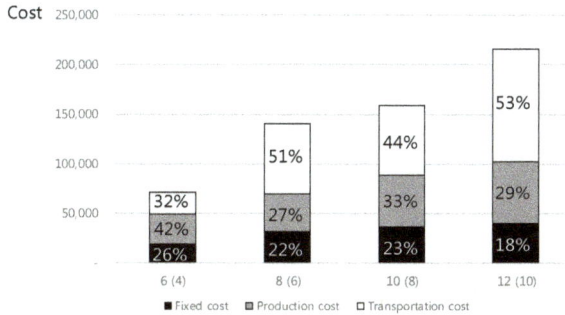

Figure 5. Cost comparison based on change in problem size.

With a demand of 100, it is observed that the total cost increases as the size of the problem increases, as shown in Figure 5. The greater the number of processes required to complete the product, the higher the total cost, with a significant increase in transportation costs. Also, it is observed that fixed costs account for a smaller proportion of total cost as the problem increases in size. Therefore, efforts should be made to reduce the number of processes required for personalized products during design.

Next, the number of processes was fixed at eight and network-connected factories at ten, to investigate the effect of a change in capacity. Figure 6a,b show the structure of the supply chain that is designed for 20% less and 30% more capacity respectively. If capacity is small, more factories are needed for each process, which leads to a complicated supply chain design and an increase in cost. One interesting fact is that the supply chain or network of nodes with a larger capacity is not a subset of the supply chain with a lower capacity. Therefore, it should be emphasized that the dynamic supply chain design should be based on real-time data, including demand and available capacity.

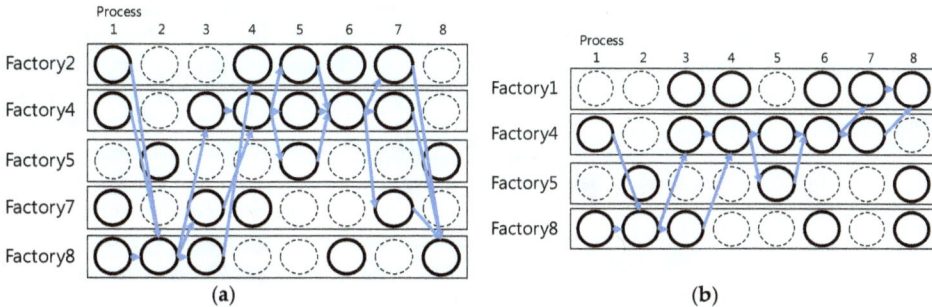

(a) (b)

Figure 6. Dynamic supply chain design. (**a**) 20% less capacity; (**b**) 30% more capacity.

The cost structures of the experiments with factory capacities that vary from −20% to 30% are shown in Figure 7. If the capacities of the factories located in the network become smaller, it becomes impossible to allocate products to less expensive factories, which increases the total cost. Because a given product can be produced in a smaller number of factories at a lower cost, the fixed cost and production cost are somewhat lower, and the transportation cost decreases significantly. For example, when we have 20% less capacity, the percentage of transportation costs is 53%, while the portion of transportation cost decreases to 42% with 20% more capacity.

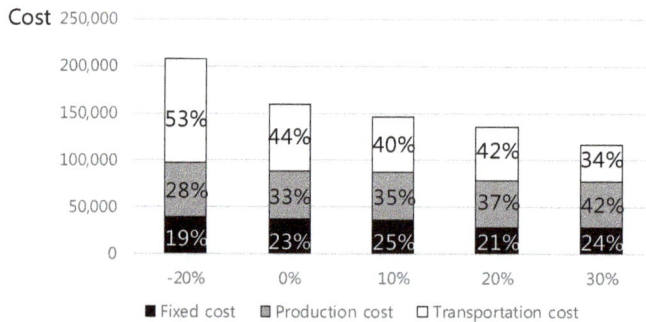

Figure 7. Cost comparison based on capacity change.

5. Conclusions

The development of technology is changing industries and enterprises. In contrast to traditional manufacturing practices, in a connected smart factory environment, advanced IT technologies must be integrated with manufacturing processes tailored to the production and delivery of personalized products. Since a factory cannot meet the personalized demand of all customers, a new concept of supply chain management is needed.

In this study, we investigated a smart supply chain with additive manufacturing that enables connected smart factories to communicate with each other in real time, to share data, and to make optimal decisions to support consumers through the cloud. First, we reviewed the literature related to the concept of smart supply chains, and presented six types of flexibility focusing on smart supply chains with additive manufacturing where processes are interchangeable, in order to support personalized customer requirements. These are design, product, process, supply chain, collaboration, and strategic flexibilities. Through a literature review and classification, we confirmed that a smart supply chain, especially with additive manufacturing, could provide more flexibility and opportunities in the face of the new manufacturing paradigms associated with personalized production. Also, flexibility is multi-dimensional [56] and smart supply chains can be flexible in different ways. Although the scope of this study does not cover the more comprehensive meaning of the smart supply chain, cloud-based networked manufacturing like FaaS can be one of the approaches for smart supply chains to flexibly meet individual requirements.

Next, we proposed a planning framework for dynamically designing and operating a smart supply chain, and formulated mathematical models for each stage, assuming a situation involving interchangeable factories and processes. Numerical experiments were conducted to compare the results of the proposed model under selected scenarios. Experimental results demonstrated the importance of dynamic supply chain design and operation with real time data, and showed that transportation costs vary widely . These proposed processes and results show that the smart supply chain is different from the traditional supply chain for mass production. First, in the traditional setting, the supply chain is designed as a long-term strategy [11,12]; in our model however, the supply chain is determined as a short-term plan. Second, in traditional supply chains, supply chain design is rather static and is not affected by individual orders. However, in our model, the supply chain is designed dynamically, based on the order received and available capacity at a given point of time; SMEs can be competitive with a dynamic network [13,14]. Third, in the smart supply chain, each customer can be involved in the supply chain design process for his or her own product [7].; this is not true of the traditional supply chain. Finally, dynamic supply chain design and operations can proactively handle the changes and uncertainties in the environments, in line with recent approaches to strategic flexibility [47].

Since the new manufacturing environment studied has not been fully implemented yet, we need to consider additional constraints on the actual environment and expand the model accordingly. In this study, the optimization models for dynamic supply chain design and operations were developed separately. In future research, we will develop a mathematical model that integrates the design and operation phase, and compares the results and performance with this study. Future studies will also take into account the diverse uncertainties that occur in environments where multiple factories are shared. Also, in future research, other relevant factors, such as quality of product, compatibility and pricing strategies, can be considered.

Acknowledgments: This work was supported by Institute for Information & communications Technology Promotion(IITP) grant funded by the Korea government(MSIT) (No. 2015-0-00366, Development of Open FaaS IoT Service Platform for Mass Personalization).

Author Contributions: Byung Do Chung and Sung Il Kim conceived and designed the experiments; Byung Do Chung and Sung Il Kim performed the experiments; Byung Do Chung and Jun Seop Lee analyzed the data; Byung Do Chung wrote the paper.

Conflicts of Interest: The authors declare no conflict of interest.

References

1. Da Silveira, G.; Borenstein, D.; Fogliatto, F.S. Mass customization: Literature review and research directions. *Int. J. Prod. Econ.* **2001**, *72*, 1–13. [CrossRef]
2. Jasti, N.V.K.; Kodali, R. Lean production: Literature review and trends. *Int. J. Prod. Res.* **2015**, *53*, 867–885. [CrossRef]
3. HU, S.J. Evolving paradigms of manufacturing: From mass production to mass customization and personalization. *Procedia CIRP* **2013**, *7*, 3–8. [CrossRef]
4. Petrick, I.J.; Simpson, T.W. 3D printing disrupts manufacturing: How economies of one create new rules of competition. *Res. Technol. Manag.* **2013**, *56*, 12–16. [CrossRef]
5. Lu, Y.; Wang, H.; Xu, X. ManuService ontology: A product data model for service-oriented business interactions in a cloud manufacturing environment. *J. Intell. Manuf.* **2016**, 1–18. [CrossRef]
6. D'aveni, R.A. 3-D printing will change the world. *Harv. Bus. Rev.* **2013**, *91*, 34–35.
7. Kagermann, H.; Helbig, J.; Hellinger, A.; Wahlster, W. *Recommendations for Implementing the Strategic Initiative INDUSTRIE 4.0: Securing the Future of German Manufacturing Industry*; final report of the Industrie 4.0 Working Group; Forschungsunion: Berlin, Germany, 2013.
8. Bughin, J.; Chui, M.; Manyika, J. Clouds, big data, and smart assets: Ten tech-enabled business trends to watch. *McKinsey Q.* **2010**, *56*, 75–86.
9. Meier, M.; Seidelmann, J.; Mezgár, I. ManuCloud: The next-generation manufacturing as a service environment. *ERCIM News* **2010**, *83*, 33–34.
10. Kim, H.-J. Bounds for parallel machine scheduling with predefined parts of jobs and setup time. *Ann. Oper. Res.* **2018**, *261*, 401–412. [CrossRef]
11. Chopra, S.; Meindl, P. Supply chain management. Strategy, planning & operation. In *Das Summa Summarum des Manage*; Gabler: Wiesbaden, Germany, 2007; pp. 265–275.
12. Ravindran, A.R.; Warsing, D.P., Jr. *Supply Chain Engineering: Models and Applications*; CRC Press: Boca Raton, FL, USA, 2012.
13. Noori, H.; Lee, W.B. Factory-on-demand and smart supply chains: The next challenge. *Int. J. Manuf. Technol. Manag.* **2002**, *4*, 372–383. [CrossRef]
14. Noori, H.; Lee, W.B. Dispersed network manufacturing: Adapting SMEs to compete on the global scale. *Int. J. Manuf. Technol. Manag.* **2006**, *17*, 1022–1041. [CrossRef]
15. Camarinha-Matos, L.M.; Afsarmanesh, H. The emerging discipline of collaborative networks. In *Working Conference on Virtual Enterprises*; Springer: Boston, MA, USA, 2004; pp. 3–16.
16. Camarinha-Matos, L.M.; Afsarmanesh, H. Collaborative networks: A new scientific discipline. *J. Intell. Manuf.* **2005**, *16*, 439–452. [CrossRef]
17. Gunasekaran, A.; Lai, K.-H.; Cheng, T.E. Responsive supply chain: A competitive strategy in a networked economy. *Omega* **2008**, *36*, 549–564. [CrossRef]
18. Kühnle, H. (Ed.) *Distributed Manufacturing: Paradigm, Concepts, Solutions and Examples*; Springer: New York, NY, USA, 2009.

19. Gaynor, M.; Moulton, S.L.; Welsh, M.; LaCombe, E.; Rowan, A.; Wynne, J. Integrating wireless sensor networks with the grid. *IEEE Internet Comput.* **2004**, *8*, 32–39. [CrossRef]
20. Bendavid, Y.; Cassivi, L. Bridging the gap between RFID/EPC concepts, technological requirements and supply chain e-business processes. *J. Theor. Appl. J. Electron. Commer. Res.* **2010**, *5*, 1–16. [CrossRef]
21. Bendavid, Y.; Cassivi, L. A 'living laboratory'environment for exploring innovative RFID-enabled supply chain management models. *Int. J. Prod. Dev.* **2012**, *17*, 94–118. [CrossRef]
22. Ivanov, D.; Sokolov, B. The inter-disciplinary modelling of supply chains in the context of collaborative multi-structural cyber-physical networks. *J. Intell. Manuf. Technol. Manag.* **2012**, *23*, 976–997. [CrossRef]
23. Butner, K. The smarter supply chain of the future. *Strategy Leadersh.* **2010**, *38*, 22–31. [CrossRef]
24. Wu, L.; Yue, X.; Jin, A.; Yen, D.C. Smart supply chain management: A review and implications for future research. *Int. J. Health Policy Manag.* **2016**, *27*, 395–417. [CrossRef]
25. Wong, K.V.; Hernandez, A. A review of additive manufacturing. *ISRN Mech. Eng.* **2012**. [CrossRef]
26. Berman, B. 3-D printing: The new industrial revolution. *Bus. Horiz.* **2012**, *55*, 155–162. [CrossRef]
27. Lu, L.; Sharf, A.; Zhao, H.; Wei, Y.; Fan, Q.; Chen, X.; Savoye, Y.; Tu, C.; Cohen-Or, D.; Chen, B. Build-to-last: Strength to weight 3D printed objects. *ACM Trans. Graph. (TOG)* **2014**, *33*, 97. [CrossRef]
28. Qin, Z.; Compton, B.G.; Lewis, J.A.; Buehler, M.J. Structural optimization of 3D-printed synthetic spider webs for high strength. *Nat. Commun.* **2015**, *6*, 7038. [CrossRef] [PubMed]
29. Beyer, C. Strategic implications of current trends in additive manufacturing. *J. Manuf. Sci. Eng.* **2014**, *136*, 064701. [CrossRef]
30. Zegard, T.; Paulino, G.H. Bridging topology optimization and additive manufacturing. *Struct. Multidiscip. Optim.* **2016**, *53*, 175–192. [CrossRef]
31. Berman, B. Should your firm adopt a mass customization strategy? *Bus. Horiz.* **2002**, *45*, 51–60. [CrossRef]
32. Insights, M.M. Layer-by-Layer: Opportunities in 3D Printing Technology Trends, Growth Drivers and the Emergence of Innovative Applications in 3D Printing. 2013. Available online: http://www.marsdd.com/wp-content/uploads/2014/04/MAR-CLT6965_3D-Printing_White_paper.pdf (accessed on 17 March 2018).
33. Campbell, T.; Williams, C.; Ivanova, O.; Garrett, B. Could 3D printing change the world. In *Technologies, Potential, and Implications of Additive Manufacturing*; Atlantic Council: Washington, DC, USA, 2011.
34. Bassoli, E.; Gatto, A.; Iuliano, L.; Grazia Violante, M. 3D printing technique applied to rapid casting. *Rapid Prototyp. J.* **2007**, *13*, 148–155. [CrossRef]
35. Sasson, A.; Johnson, J.C. The 3D printing order: Variability, supercenters and supply chain reconfigurations. *Int. J. Phys. Distrib. Logist. Manag.* **2016**, *46*, 82–94. [CrossRef]
36. Khan, A.; Turowski, K. A survey of current challenges in manufacturing industry and preparation for industry 4.0. In *Proceedings of the First International Scientific Conference "Intelligent Information Technologies for Industry" (IITI'16)*; Springer: Cham, Switzerland, 2016; pp. 15–26.
37. Cotteleer, M.; Joyce, J. 3D opportunity: Additive manufacturing paths to performance, innovation, and growth. *Deloitte Rev.* **2014**, *14*, 5–19.
38. Liu, P.; Huang, S.H.; Mokasdar, A.; Zhou, H.; Hou, L. The impact of additive manufacturing in the aircraft spare parts supply chain: Supply chain operation reference (scor) model based analysis. *Prod. Plan. Control* **2014**, *25*, 1169–1181. [CrossRef]
39. Zhang, Y.; Huang, G.Q.; Qu, T.; Sun, S. Real-time work-in-progress management for ubiquitous manufacturing environment. In *Cloud Manufacturing*; Springer: London, UK, 2013; pp. 193–216.
40. Moghaddam, M.; Nof, S.Y. Real-time optimization and control mechanisms for collaborative demand and capacity sharing. *Int. J. Prod. Econ.* **2016**, *171*, 495–506. [CrossRef]
41. Hsieh, C.-C.; Wu, C.-H. Capacity allocation, ordering, and pricing decisions in a supply chain with demand and supply uncertainties. *Eur. J. Oper. Res.* **2008**, *184*, 667–684. [CrossRef]
42. Mohr, S.; Khan, O. 3D Printing and Supply chains of the Future. Innovations and Strategies for Logistics and Supply Chains. In Proceedings of the Hamburg International Conference of Logistics, Hamburg, Germany, 24–25 September 2015.
43. Seok, H.; Nof, S.Y. Collaborative capacity sharing among manufacturers on the same supply network horizontal layer for sustainable and balanced returns. *Int. J. Prod. Res.* **2014**, *52*, 1622–1643. [CrossRef]
44. Chen, F. Information sharing and supply chain coordination. *Handb. Oper. Res. Manag. Sci.* **2003**, *11*, 341–421.
45. Chan, H.K.; Chan, F.T. Effect of information sharing in supply chains with flexibility. *Int. J. Prod. Res.* **2009**, *47*, 213–232. [CrossRef]

46. Jun, C.; Wei, M.Y. The research of supply chain information collaboration based on cloud computing. *Procedia Environ. Sci.* **2011**, *10*, 875–880. [CrossRef]

47. Brozovic, D. Strategic flexibility: A review of the literature. *Int. J. Manag. Rev.* **2018**, *20*, 3–31. [CrossRef]

48. Aaker, D.A.; Mascarenhas, B. The need for strategic flexibility. *J. Bus. Strategy* **1984**, *5*, 74–82. [CrossRef]

49. Sanchez, R. Strategic flexibility in product competition. *Strat. Manag. J.* **1995**, *16*, 135–159. [CrossRef]

50. Verdú-Jover, A.J.; Lloréns-Montes, F.J.; García-Morales, V.J. Environment–flexibility coalignment and performance: An analysis in large versus small firms. *J. Small Bus. Manag.* **2006**, *44*, 334–349. [CrossRef]

51. Scott, A.; Harrison, T.P. Additive manufacturing in an end-to-end supply chain setting. *3D Print. Addit. Manuf.* **2015**, *2*, 65–77. [CrossRef]

52. Shirodkar, S.; Kempf, K. Supply chain collaboration through shared capacity models. *Interfaces* **2006**, *36*, 420–432. [CrossRef]

53. Barz, A.; Buer, T.; Haasis, H.-D. A study on the effects of additive manufacturing on the structure of supply networks. *IFAC-PapersOnLine* **2016**, *49*, 72–77. [CrossRef]

54. Emelogu, A.; Marufuzzaman, M.; Thompson, S.M.; Shamsaei, N.; Bian, L. Additive manufacturing of biomedical implants: A feasibility assessment via supply-chain cost analysis. *Addit. Manuf.* **2016**, *11*, 97–113. [CrossRef]

55. Ivanov, D.; Dolgui, A.; Sokolov, B.; Werner, F.; Ivanova, M. A dynamic model and an algorithm for short-term supply chain scheduling in the smart factory industry 4.0. *Int. J. Prod. Res.* **2016**, *54*, 386–402. [CrossRef]

56. Zukin, M.; Dalcol, P.R. Manufacturing flexibility: Assessing managerial perception and utilization. *Int. J. Flex. Manuf. Syst.* **2000**, *12*, 5–23. [CrossRef]

![applied sciences logo] *applied sciences*

MDPI

Article

The Role of Managerial Commitment and TPM Implementation Strategies in Productivity Benefits

José Roberto Díaz-Reza [1], Jorge Luis García-Alcaraz [2,*], Liliana Avelar-Sosa [2], José Roberto Mendoza-Fong [1], Juan Carlos Sáenz Diez-Muro [3] and Julio Blanco-Fernández [4]

[1] Department of Electrical and Computer Engineering, Universidad Autónoma de Ciudad Juárez, Ciudad Juarez 32310, Mexico; al164440@alumnos.uacj.mx (J.R.D.-R.); al164438@alumnos.uacj.mx (J.R.M.-F.)
[2] Department of Industrial Engineering and Manufacturing, Universidad Autónoma de Ciudad Juárez, Juarez 32310, Mexico; liliana.avelar@uacj.mx
[3] Department of Electrical Engineering, Universidad de La Rioja, 26006 Logroño, Spain; juan-carlos.saenz-diez@unirioja.es
[4] Department of Mechanical Engineering, Universidad de La Rioja, 26006 Logroño, Spain; julio.blanco@unirioja.es
* Correspondence: jorge.garcia@uacj.mx; Tel.: +52-656-688-4843 (ext. 5433)

Received: 28 June 2018; Accepted: 10 July 2018; Published: 16 July 2018

Featured Application: The findings in this research allow managers to know quantitatively the importance of preventive maintenance in an adequate performance of total productive maintenance and productivity benefits in a production system, allowing them pay attention on those activities that are more important.

Abstract: The present research proposes a structural equation model to integrate four latent variables: managerial commitment, preventive maintenance, total productive maintenance, and productivity benefits. In addition, these variables are related through six research hypotheses that are validated using collected data from 368 surveys administered in the Mexican manufacturing industry. Consequently, the model is evaluated using partial least squares. The results show that managerial commitment is critical to achieve productivity benefits, while preventive maintenance is indispensable to total preventive maintenance. These results may encourage company managers to focus on managerial commitment and implement preventive maintenance programs to guarantee the success of total productive maintenance.

Keywords: TPM; implementation; managerial commitment; productivity benefits

---

1. Introduction

In current industrial scenarios, waste in production processes is frequent, usually the result of the lack of skills of both operators and maintenance staff, not enough machinery available, and issues with work tools [1]. Other types of waste include machinery downtime, not utilizing talents, damaged machinery, and rejected parts, among others [2].

To increase competitiveness, manufacturing companies seek to reduce the activities that add no value to a product but generate cost, and in this sense, one way of reducing waste is to adopt a lean approach. The lean manufacturing (LM) approach aims to reduce the amount of non–value-added activities in the production process, although it also has reported benefits at the administrative level [3]. LM relies on several tools to achieve its goal, and total productive maintenance (TPM) is one of the most important, because it helps companies to minimize waste, such as damaged machinery and unplanned work, and it encourages the development of production plans that prevent machine overload [4].

Furthermore, TPM can be defined as an approach that rapidly improves production processes through employee involvement and empowerment [5]. Nowadays, in the restless and uncertain global business environment, well-managed organizations strive to improve their capabilities by operating profitably; in other words, TPM is a tool that, if correctly implemented, can help businesses reach this goal [6].

Currently, TPM is a successful tested LM tool for planning the maintenance of organizational activities, which involve operators and maintenance staff working together as a team [7]. In this sense, TPM is associated with human resources, and it integrates equipment maintenance in the production process to increase machine availability, as well as adding commercial value to the organization [8].

Also, TPM aims to keep production equipment in proper working condition to prevent breakdowns that eventually delay the production process or make unsafe workplaces, thus TPM is one of the main operational activities in quality management systems [8]. Additionally, TPM emphasizes proactive and preventive maintenance to maximize operational production machinery efficiency and decrease the roles of the production and maintenance departments by empowering operators [8]; it improves organizational competitiveness and comprises a powerful and structured approach to changing employee mentality and, consequently, the organizational culture. This also includes employee involvement at all hierarchical levels in all company departments [9]. In addition, it involves member alignment to improve corrective actions associated with safety [7].

Similarly, TPM not only focuses on machine efficiency, but also is an opportune area for continuous improvement, looking for an ideal relationship between people and machines. In other words, successful organizations must be supported by effective and efficient maintenance plans as a competitive strategy [10]. However, production systems are not the only places where TPM can be applied, because other service-based systems need to be in optimal operating condition as well, such as medical equipment and instruments [11].

Because of the benefits that TPM brings to the industrial environment, one of the main academic and industrial concerns is to find its critical success factors (CSFs). Identifying CSFs allows company managers and administrators to prioritize the activities that ensure TPM success. Also, many studies have reported the key activities involved in these TPM factors [12–15] as well as the obtained benefits [16,17]. However, the relationship between these success factors and company benefits have not been clearly defined.

In addition, it is observed that there is an academic and industrial interest to identify the CSFs of TPM, as well as the activities that integrate them [12–15,18,19]. Similarly, there are several reports associated with the obtained benefits from proper TPM implementation in production systems [16,17,20]; however, the problem is that there are not enough studies that link CSF with obtained benefits, and as a result, it has not been determined which CSFs are crucial for obtaining specific desired benefits.

Kamath and Rodrigues [21] state that the CSFs may be different from one industrial sector to another and that findings cannot be generalized. For example, Chlebus, et al. [22] report on activities associated with TPM in a mine, and Ahuja and Khamba [23] report on manufacturing industries in India. In Mexico there are currently 5518 maquiladora industries with manufacturing and export services, and in Chihuahua state there are 510, and these amount to 9.25% of the national total. Of those 510 in Chihuahua state, Ciudad Juárez has 332, which is 65.10% of the state, directly employing 268,761 workers [24]. Those maquiladora companies are characterized by a high technological level that requires a lot of maintenance services, and there are not enough studies indicating which are the CSFs of TPM and the benefits gained in that industrial sector.

Because the relationship between CSFs for TPM and gained benefits is currently an interesting research area, this paper presents a structural equation model that associates three CSFs: managerial commitment, TPM implementation, and PM implementation, which are related to productivity benefits. In addition, findings in this paper will help managers identify the most significant activities to have a successful TPM implementation and guarantee its benefits.

The purpose of this research is to quantify, through a structural equation model, the impact of managerial commitment, the implementation process, and the plans and programs executed as CSFs for TPM on productivity benefits gained, and it is validated using information from the maquiladora industry in Mexico.

The rest of this paper is organized as follows: Section 2 provides a brief introduction to concepts related to managerial importance in TPM implementation and its benefits, Section 3 presents the research hypotheses that link the studied variables, Section 4 describes the research methodology, Section 5 reports the research findings, and, finally, Section 6 shows the research conclusions and industrial implications.

2. Maintenance and Concepts

2.1. Critical Success Factors of TPM

CSFs are relevant performance areas that help companies reach desired goals (e.g., TPM goals) [25]. The literature on TPM addresses a large number of CSFs for TPM implementation. For instance, Park and Han [17] consider that a key factor is employee involvement, since the true power of TPM is using employee knowledge and experience to generate ideas to achieve the desired goals and objectives. On the other hand, Ng, Goh and Eze [12] found that human resources elements along with managerial commitment, employee involvement, education, and training are fundamental in TPM.

In their research, Piechnicki, et al. [26] identified a set of critical TPM success factors and grouped them into eight categories: education and training, teamwork, planning and preparation, senior managerial commitment, resistance to change, change of culture, employee involvement, monitoring results, and effective communication; as can be seen, there are a lot of CSF human resources for TPM. However, Hernández Gómez et al. [27] classified three categories for CSF: strategic planning, technical aspects, and human resources development. That list supports the importance of human resources in TPM success.

In a recent study, Gómez, Toledo, Prado, and Morales [14] performed a factor analysis to evaluate 31 key activities in the TPM implementation process, and the results revealed nine critical success factors: strategic alignment, continuous improvement practices, plant distribution, autonomous maintenance, equipment alignment, employee and supplier involvement, cutting-edge technology, technology development, and communication regarding the TPM development process. This research reveals that top managers must be highly committed to TPM, since they provide the necessary implementation resources. Similarly, operators are equally necessary to detect and prevent errors before failures occur; this approach to equipment maintenance is called preventive maintenance, which is another critical success factor of TPM.

2.2. Preventive Maintenance

Preventive maintenance (PM) is defined as a set of activities performed at certain times in a planning horizon to extend the equipment life cycle and keep it in satisfactory working condition, and to increase overall system reliability and availability [28]. Such activities are a part of maintenance programs and attempt to minimize the risk of unplanned equipment downtime. According to its nature, PM includes inspection, cleaning, lubrication, adjustment, alignment, and component replacement for machinery and tools in a production process [28].

Those tasks play a vital role in any production process, as they preserve equipment operating under desired long-term specifications [29]. Also, PM has a positive impact on cost, quality, and delivery performance [30], because it minimizes quality costs by keeping the equipment in the best working condition through proper maintenance programs that guarantee a high rate of compliant products [31].

Currently there are studies that reveal the importance of PM in production systems. For example, Eti et al. [32] make an association between the cost of PM and increased reliability of machinery and equipment; Anis et al. [33] propose a plan to relate PM program activities with batch size for production lines to minimize the time lost due to stoppages associated with maintenance. Shrivastava et al. [34] state that product quality is not the responsibility of one department, but preventive maintenance and calibration and adjustments in machinery and equipment are essential to guarantee products with specifications required for clients. Finally, given that PM requires stopping machines and equipment continuously, Fumagalli et al. [35] recommend adequate orchestration of maintenance plans with product delivery commitments in the production system.

All of the above show the importance of PM as a work culture focus for conservation of machinery, and this has industrial and academic interest.

2.3. TPM Benefits

TPM critical success factors guarantee important benefits. According to Willmott and McCarthy [16], they provide opportunities to develop suitable strategic plans for the company's capabilities, infrastructure, and human resources to support integration between the operation and maintenance departments, and improve business relationships with customers and suppliers. Similarly, Ma et al. [36] reported that TPM increases production system efficiency, brings social benefits, and promotes management systems in production departments. On the other hand, Gupta, Vardhan, and Haque [13] mention that TPM increases employee morale and skills, improves the use of technology, and enhances equipment working conditions and customer satisfaction. Additionally, successful industries that implement TPM increase their overall equipment efficiency (OEE) by 30%, and some have increased OEE by 95%.

In addition, Ng et al. [37] found that TPM minimizes machine downtime, increases employee motivation, and minimizes accidents. Likewise, Rodrigues and Hatakeyama [38] argue that companies can increase daily production rates with their current productive capacity and workforce if they implement TPM strategies; however, the researchers also claim that many companies tend to spend little time on equipment maintenance and struggle to empower operators to help maintain their equipment.

3. Hypotheses and Literature Review

The goal of TPM is to maximize equipment effectiveness by continuously improving availability and preventing failures, but this cannot be achieved without managerial support [39] and employee involvement. Therefore, participation by all employees promotes a preventive maintenance approach based on motivation management and voluntary small group participation [40]. Also, TPM concepts entail a long-term commitment to planning, especially from senior managers, and usually initiates as a top-down exercise, but it can only be successfully implemented as "bottom-up" participation [7]. From this perspective, the operator commitment, performance, and morale reflect the managerial commitment to TPM [41]. In conclusion, managerial commitment is the most essential factor in TPM implementation [39], which is why the first hypothesis in this research can be proposed as follows:

Hypothesis 1. *Managerial commitment has a positive and direct effect on TPM implementation.*

TPM initiates with senior managers, since they have the power to implement the necessary organizational changes and formulate the required plans, strategies, and policies, which must be aligned with corporate objectives. Also, the maintenance manager is responsible for the effective performance of TPM activities and setting the basic policies and objectives to be reached through a carefully designed plan [39]. Similarly, senior managers should remove the obstacles that interfere with TPM plans and implementation, and they must make sure that such plans are aligned with the company's short- and long-term goals [42]. In other words, because TPM is based on error prevention, it is important to rely on effective preventive maintenance techniques. PM must be aligned with managerial preferences and priorities in a production system, since that guarantees workers' acceptance, and it always must be focused on preventing accidents or failures [43]. Therefore, the second hypothesis in this research can be presented as follows:

Hypothesis 2. *Managerial commitment has a positive and direct effect on PM implementation.*

The main TPM characteristics are economic efficiency, preventive maintenance, improved maintenance capacity, and employee involvement [44]. As Nakajima [45] argues, the main goal in the TPM implementation stage is to increase equipment efficiency through specific techniques, including autonomous maintenance, where employees improve their skills and help to maintain their equipment. From this perspective, it is claimed that PM is part of the organizational culture and an essential component of TPM; that is, there must be an error and failure prevention program to ensure TPM. Therefore, the third hypothesis in this research can be proposed as follows:

Hypothesis 3. *PM implementation has a positive and direct impact on TPM implementation.*

As Nakajima [46] points out, the main TPM objectives are to improve productivity, increase management efficiency, and eliminate the six types of production waste. In addition, managerial commitment, along with clear and common goals and visible results, creates a responsibility shared by a team and leads to fewer interruptions and higher reliability levels [47]. Therefore, TPM relies on long-term managerial commitment to increase equipment efficiency and efficacy in order to offer the expected benefits [9]. As a result, the fourth hypothesis in this research can be presented as follows:

Hypothesis 4. *Managerial commitment has a positive direct effect on productivity benefits.*

TPM implementation can minimize waste and redundant work, and increase company profitability and image, which guarantee competitiveness for a company [39]. Likewise, TPM improves OEE, productivity, safety, and quality, and minimizes equipment life cycle costs [48]. In addition, implementing a TPM approach leads to increased efficiency and work quality; minimizes customer complaints, accidents, and internal waste [49]; and improves delivery performance, stock turnover and employee morale, productivity, and performance [50]. Therefore, the fifth hypothesis in this research is proposed as follows:

Hypothesis 5. *TPM implementation has a positive direct effect on productivity benefits.*

PM is a scheduled activity usually initiated under certain statistical parameters (e.g., average time, usage) that determine when maintenance actions are necessary before entering the risk zone (where the probability of random equipment component failure increases) [51]. In addition, PM must help prevent unplanned equipment downtime and maximize system availability by keeping equipment in proper working condition and improving its availability. PM includes scheduled tasks, such as supervision of hardware replacement and control before the equipment ends its life cycle. Due to these PM activities, there is a direct impact on the production system [52]; the sixth hypothesis in this research can be presented as follows:

Hypothesis 6. *PM implementation has a positive direct effect on productivity benefits.*

Figure 1 illustrates the established relationships between the studied variables relating CSFs for TPM and gained benefits.

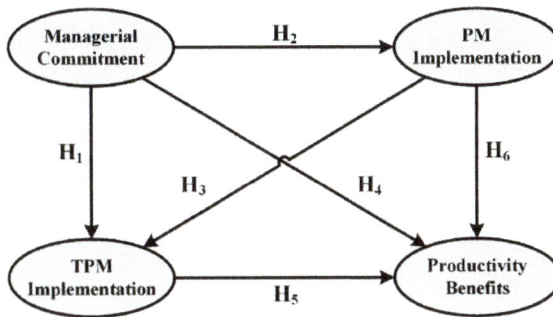

Figure 1. Initial model. TPM, total productive maintenance; PM, preventive maintenance.

4. Materials and Methods

4.1. Stage 1: Survey Design

In order to validate the model in Figure 1, a survey instrument to gather data was designed. As a matter of fact, designing the survey included reviewing works similar to this research to identify which activities (items) integrate each latent variable (e.g., CSF for TPM and productivity benefits). The previous studies were collected from multiple databases, including ScienceDirect, Springer, and IEEE, among others. This literature review represented validation of the survey's rationale [53] and Table 1 illustrates the CSF for TPM (activities) and productivity benefits gained that were identified.

Table 1. Total productive maintenance (TPM) critical success factors and productivity benefits.

Latent Variable	Items
preventive maintenance (PM) Implementation [51,54]	preventive maintenance as a quality strategy. maintenance department committed to prevention and operator support. report the maintenance actions performed on the equipment. disclose statistics of the maintenance records. easy access to equipment maintenance records. record the quality generated by the equipment. identify causes of machine failures and report the statistics.
TPM implementation [4,26,27]	proper education and training of maintenance staff. follow-up and control of the maintenance program. commitment from managers and maintenance staff. managerial leadership in TPM execution. leadership from production and engineering departments in TPM execution. maintenance staff leadership in TPM execution. communication between production and maintenance departments. knowledge of critical machine systems. TPM focused on the life cycle of machine systems, parts, and components. purchase of machines and equipment based on TPM.
managerial commitment [26,27,39]	department leaders embrace their TPM responsibilities. top managers lead TPM execution. meetings are held between production and maintenance departments. top managers promote employee participation and encourage preservation of the work team. top managers develop and communicate a quality- and maintenance-centered vision. top managers are directly involved in maintenance projects.
productivity benefits [15,37]	elimination of productivity losses. increased equipment reliability and availability. reduction of maintenance costs. improved final product quality. decreased spare parts inventory costs. improved corporate technology. improved response to market changes. development of corporate competitive skills

Next, the identified activities were used to develop a preliminary survey integrated by three sections: sociodemographic information, activities of critical success factors for TPM and PM, and productivity benefits.

4.2. Stage 2: Survey Administration

The final questionnaire was given to TPM practitioners in the Mexican manufacturing industry: senior managers, engineers, technicians, supervisors, and operators. In order to select the sample, first a stratified sampling technique was followed; that is, companies with fully consolidated maintenance programs (10 years of implementation or more) that could corroborate such information through equipment maintenance records were identified and considered.

The survey was answered with a five-point Likert scale, as seen in Table 2, through face-to-face interviews. The lowest value on the scale (1) indicated that an activity was never performed or a productivity benefit was never obtained, whereas the highest value (5) indicated that an activity was always performed or a productivity benefit was always obtained. Also, experienced personnel who were interviewed recommended other possible responders, then the snowball sampling technique was implemented.

Table 2. Survey scale.

Value	1	2	3	4	5
interpretation	never	rarely	regularly	frequently	always

4.3. Stage 3: Data Capture and Screening

The collected data were registered in a database using SPSS 24®; the columns represented the survey items and the rows cases or questionnaires. Then, the database was screened by performing the following operations:

- The standard deviation was calculated; if the value was lower than 0.5, that case was removed, since all the items had a similar value [55].
- The missing values were identified; if a questionnaire had 10% or more missing values, it was discarded [56]. On the other hand, for questionnaires that had less than 10% of missing values, such values were replaced with the median in the item [57].
- Outliers were identified by standardizing each item; extreme or atypical observations with an absolute standardized value greater than 4 [58,59] were replaced by a median value in the item.

4.4. Stage 4: Survey Validation

Once the data were evaluated, the 4 latent variables were tested by estimating the following indices proposed by Kock [60]:

- R-squared and adjusted R-squared for the predictive validity of the survey from a parametric perspective; only values over 0.2 were acceptable.
- Q-squared for the predictive validity of the survey from a nonparametric perspective; acceptable Q-squared values, and their R-square values, must be greater than 0.
- Cronbach's alpha and compound reliability index for internal variability of the latent variables; internal validity can be estimated based on the variance or correlation index between the items of a latent variable [61], and acceptable values must be greater than 0.7.
- Average variance extracted (AVE) for the convergent validity of the items in the latent variables; acceptable values must be greater than 0.5.
- Average block variance inflation factor (VIF) and average full collinearity VIF (AFVIF) for the collinearity of the items in the latent variables; acceptable values must be less than 3.3.

4.5. Stage 5: Structural Equation Model

The 4 latent variables were integrated into a structural equation model (SEM), as illustrated in Figure 1, with 6 research hypotheses. In addition, the SEM was tested using partial least squares (PLS), which is widely accepted in multiple disciplines [62]. Also, the greatest advantage of an SEM is its ability to model and illustrate, at the same time, the direct and indirect interrelations between multiple dependent and independent latent variables, because the latent variables have different roles, as dependent and independent, in this research. Likewise, SEMs are reliable even using nonnormal data, small samples, or ordinal data [63].

The research hypotheses shown in Figure 1 were tested using WarpPLS v.6.0® software (ScriptWarp Systems, Laredo, TX, USA, 2017), which is based on PLS; it is widely recommended by Kock [64], and some PLS applications can be found, for instance, in Midiala Oropesa, et al. [65], who modeled the effects of Kaizen in an industrial context, or in García-Alcaraz, et al. [66], who modeled the effects of Just in Time (JIT) in the manufacturing industry.

The model was tested with a 95% reliability level, implying that the *p*-values of the parameters had to be lower than 0.05. Before interpreting the SEM, 6 efficiency models and quality indices were calculated, proposed by Kock [60]:

- Average path coefficient (APC): statistically validates the hypotheses in a generalized way. The *p*-value must be less than 0.05.
- Average R-squared (ARS) and average adjusted R-squared (AARS): measure the model's predictive validity. Acceptable *p*-values for ARS and AARS must be less than 0.05. The null hypotheses to be tested are APC = 0 and ARS = 0 against the alternative hypotheses, where APC $\neq$ 0 and ARS $\neq$ 0.
- Average variance inflation factor (AVIF) and average full collinearity VIF (AFVIF): measure the level of collinearity between the latent variables. The acceptable value must be less than 3.3.
- Tenenhaus goodness of fit (GoF): measures the explanatory power of the model. The GoF value must be greater than 0.36.

4.5.1. Direct Effects

The direct effects in the model were evaluated and are illustrated in Figure 1 by arrows directly connecting two latent variables, where each arrow represents a hypothesis. In addition, each effect has a β value and a *p*-value; β expresses dependency in standard deviations between an independent and a dependent latent variable, whereas *p*-value is for the hypothesis test where the null hypothesis is $\beta_1 = 0$, which is tested against the alternative hypothesis, $\beta_1 \neq 0$ [60]. Additionally, R^2 for the dependent variables was estimated as a coefficient that shows the amount of variance in a dependent latent variable that is explained by an independent latent variable.

4.5.2. Indirect Effects and Total Effects

In SEMs, indirect effects occur between 2 latent variables through other latent variables, known as mediators. Indirect effects also have *p*-values to determine whether they are statistically significant or not. On the other hand, total effects in a relationship are the total direct and indirect effects; total effects also have associated *p*-values.

Finally, for each effect (direct, indirect, or total), the effect size for decomposition of R-squared was estimated when 2 or more independent latent variables influenced a dependent latent variable.

4.5.3. Sensitivity Analysis

Frequently, the relationship between the latent variables is not explained enough by β values and it is necessary to know different scenarios for them. In this research, for every relationship or hypothesis in Figure 1, the probability of occurrence for 2 scenarios is analyzed when each variable occurs independently with low and high values. A third scenario represents the probability of the combination of both variables in a hypothesis when they occur simultaneously, while a fourth scenario is about the probability that the dependent variable will occur in a high or low scenario because the independent variable has occurred in a high or low scenario (a conditional probability). Since the latent variables are standardized, values greater than 1 represent high scenarios in a latent variable, while values less than 1 represent low scenarios for a latent variable.

In this research, scenarios with low values are represented by a minus sign (−) and scenarios with high values are represented by a plus sign (+). In the same way, the probability of simultaneous or simultaneous occurrence of scenarios between 2 variables (low or high) is represented by an ampersand (&). Finally, the conditional probability of occurrence of a scenario in a dependent latent variable because the scenario for the dependent variable has happened is represented by "If."

In addition, since the automotive industrial sector in this geographic context is one of the most representative in previous surveys reported by Mendoza-Fong, et al. [67], that sector is compared with other sectors to find significant differences among them, so that a model is executed for the automotive sector and a model is rejected for others, and differences in β are tested.

5. Results

5.1. Sample Description

After four months of survey administration, 368 questionnaires were collected. Table 3 lists the surveyed industrial sectors and the respondents' job positions. As can be observed, most of the sample is from the automotive industry (74 technicians, 41 operators, 32 engineers, 22 supervisors, 1 manager, and 2 other job positions).

Table 3. Industrial sector vs. job position.

Job Position	Industrial Sector						
	Aeronautics	Electrics	Automotive	Electronics	Medical	Other	Total
technician	0	21	74	29	7	11	142
operator	1	4	41	11	10	12	79
engineer	3	6	32	7	4	1	53
supervisor	1	9	22	8	2	6	48
manager	0	1	1	1	2	1	6
other	0	0	2	0	0	1	3
total	5	41	172	56	25	32	331

Also, only 331 respondents provided information. Most of the respondents are maintenance technicians or operators; together the categories represent 66.67% of the sample. Such results imply that the data collected from the survey were obtained from people directly involved in equipment maintenance.

5.2. Survey Statistical Validation

Table 4 presents the estimated coefficients or indices for the latent variables. Based on these indices, we concluded that the latent variables have enough parametric and nonparametric predictive validity, internal validity, and convergent validity. Similarly, according to the VIF values, the latent variables are free from internal collinearity problems. Consequently, because the latent variables passed the validation process, they were integrated into the model and evaluated.

Table 4. Survey validation.

Index	Managerial Commitment	TPM Implementation	PM Implementation	Productivity Benefits
R-squared	–	0.628	0.407	0.382
adj. R-squared	–	0.626	0.405	0.377
composite reliability	0.942	0.939	0.902	0.956
cronbach's alpha	0.926	0.928	0.873	0.948
average	0.729	0.608	0.569	0.732
variance inflation factor	2.560	2.804	1.982	1.499
Q-squared	–	0.630	0.409	0.381

5.3. Structural Equation Model

Table 5 shows the model fit and estimated quality indices in the model. Based on the APC, ARS, and AARS values, the model has enough predictive validity. Similarly, the VIF and AFVIF values demonstrate that the model is free from collinearity problems, whereas GoF shows that the model fits the data. According to these data, the effects between the variables can be interpreted.

Table 5. Model fit and quality indices.

Index	Value
average path coefficient (APC)	$0.368, p < 0.001$
average R-squared (ARS)	$0.472, p < 0.001$
average adjusted R-squared (AARS)	$0.470, p < 0.001$
average block VIF (AVIF); acceptable if ≤ 5, ideally ≤ 3.3	1.912
average full collinearity VIF (AFVIF); acceptable if ≤ 5, ideally ≤ 3.3	2.211
tenenhaus goodness of fit (GoF); small ≥ 0.1, medium ≥ 0.25, large ≥ 0.36	0.558

5.3.1. Direct Effects

Figure 2 presents the model results once the latent variable coefficients, model fit, and quality indices were estimated. As can be inferred from the *p*-values, all direct effects or direct relationships between the latent variables are statistically significant with a 95% reliability level. Table 6 summarizes the conclusions regarding the research hypotheses. All of the research hypotheses were accepted, since they are statistically significant.

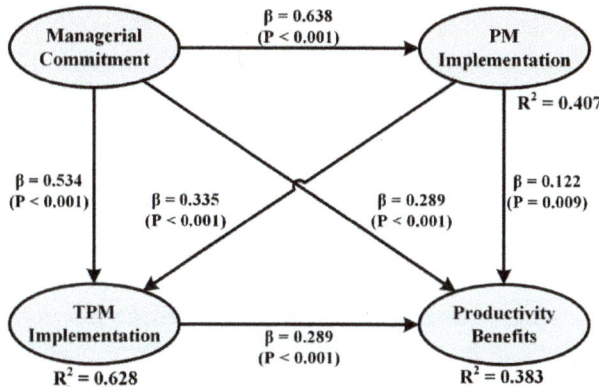

Figure 2. Evaluated model.

Table 6. Hypothesis validation.

Hypothesis	Independent Variable	Dependent Variable	β	*p*-Value	Conclusion
H1	managerial commitment	TPM implementation	0.534	<0.001	accepted
H2	managerial commitment	PM implementation	0.638	<0.001	accepted
H3	PM implementation	TPM implementation	0.335	<0.001	accepted
H4	managerial commitment	productivity benefits	0.289	<0.001	accepted
H5	TPM implementation	productivity benefits	0.289	<0.001	accepted
H6	PM implementation	productivity benefits	0.122	=0.009	accepted

Figure 2 also shows the R^2 values from the dependent latent variables. In SEMs, R^2 expresses the amount of variance in a dependent variable that is explained by one or more independent variables. For instance, the TPM implementation latent variable shows R^2 = 0.628, which is explained by managerial commitment (0.400) and PM implementation (0.228). In turn, PM implementation is explained by only one latent variable, managerial commitment, with R^2 = 0.407.

Finally, the productivity benefits latent variable is explained in 0.382 units by three latent variables: managerial commitment (0.162), PM implementation (0.057), and TPM implementation (0.164). These results imply that to ensure proper TPM implementation, it is important to perform tasks from managerial commitment, since this latent variable explains 40% of the PM implementation. Similarly, to obtain productivity benefits, companies must have managerial commitment and TPM implementation, because those variables have a higher explanatory level.

5.3.2. Indirect Effects

Figure 2 displays four indirect effects, the *p*-value for its statistical validation, and the size effect (SE) as an explanatory power: three of them occur through two segments, and one occurs through three segments. Similarly, Table 7 lists the sum of indirect effects and associates them with their corresponding *p*-values and SEs. For instance, managerial commitment is indirectly related to TPM implementation through PM implementation. This indirect relationship is statistically significant ($p < 0.001$) and has a value of 0.214, explaining 16.0% of variability.

Table 7. Indirect effects.

Dependent Variable	Independent Variable	
	Managerial Commitment	PM Implementation
TPM implementation	0.214 *p* < 0.001 ES = 0.160	-
productivity benefits	0.293 *p* < 0.001 ES = 0.165	0.097 *p* = 0.004 ES = 0.045

Likewise, the managerial commitment latent variable has an indirect effect on productivity benefits through two latent variables, TPM implementation and PM implementation; the value of this indirect effect is 0.293 and it can explain 16.5% of variability. Finally, PM implementation has an indirect effect on productivity benefits through TPM implementation with a low value of 0.097, but it is still statistically significant and can explain 4.5% of variability.

5.3.3. Total Effects

Total effects in a relationship are total direct and indirect effects. In this sense, Table 8 shows the total effects found in the model, the *p*-value associated with the statistical test, and the size effect. For instance, the total effects in the relationship between managerial commitment and productivity benefits (0.582) consists of the total direct effect (0.289) and the indirect effect (0.232).

5.3.4. Sensitivity Analysis

Table 9 presents the probabilities of occurrence independently for the latent variables analyzed by their high and low scenarios. For example, it is observed that the probability that managerial commitment is presented at a low level is only 0.158, which represents a risk for the maintenance manager, since it fully demonstrates that TPM implementation can only be possible if there is precedence for it. Similarly, the probability of having managerial commitment in a high scenario is 0.190. Interpretations of the other latent variables are performed in a similar way; in this case, the low levels represent a risk or an improvement opportunity for maintenance managers to avoid this occurrence.

Table 8. Total effects.

Dependent Variable	Independent Variable		
	Managerial Commitment	TPM Implementation	PM Implementation
TPM implementation	$0.748\ p < 0.001$ ES = 0.561	-	$0.335\ p < 0.001$ ES = 0.228
PM implementation	$0.638\ p < 0.001$ ES = 0.407	-	-
productivity benefits	$0.582\ p < 0.001$ ES = 0.326	$0.289\ p < 0.001$ ES = 0.164	$0.218\ p < 0.001$ ES = 0.102

Table 9. Scenarios and probabilities of independent occurrence.

Latent Variable	Scenario	Probability
managerial commitment	−	0.158
	+	0.190
TPM implementation	−	0.160
	+	0.166
PM implementation	−	0.190
	+	0.179
productivity benefits	−	0.155
	+	0.190

According to the previous information and based on the values of the variables in their low scenarios, one of the biggest risks is to have a failure in the PM implementation process, since there is a probability of 0.190. Consequently, managers must work hard to generate a work culture focused on preservation of production equipment, because according to the model presented in Figure 1, the success of a more complete program depends on it, as the TPM implementation. Likewise, it is observed that two latent variables have a high probability of occurrence in their high scenarios, managerial commitment and productivity benefits, with 0.190, therefore managers must focus their efforts on increasing those values.

Table 10 shows the high and low scenario combinations for the variables analyzed, where the dependent variables are presented in rows and the independent variables in columns, although each value in every relationship can be discussed. In this section, some of them are analyzed in an illustrative way. In addition, for each of the intersections, the probability of occurring simultaneously for the two variables is represented by "&" and the conditional probability of occurrence of a dependent variable because the independent variable has occurred is represented by "If." Thus, there are scenarios that are pessimistic where the two levels of the variables are low, and there are scenarios that are optimistic because the two variables are at their high level.

Table 10. Scenarios and probabilities for independent occurrence.

Dependent Variable	Scenario	Independent Variable					
		Managerial Commitment		TPM Implementation		PM Implementation	
		−	+	−	+	−	+
TPM implementation	−	& = 0.098 If = 0.621	& = 0.003 If = 0.016				
	+	& = 0.005 If = 0.034	& = 0.095 If = 0.565				
PM implementation	−	& = 0.090 If = 0.569	& = 0.008 If = 0.048	& = 0.092 If = 0.486	& = 0.003 If = 0.014		
	+	& = 0.008 If = 0.052	& = 0.084 If = 0.500	& = 0.005 If = 0.030	& = 0.087 If = 0.485		
productivity benefits	−	& = 0.065 If = 0.414	& = 0.00 If = 0.00	& = 0.076 If = 0.475	& = 0.003 If = 0.016	& = 0.065 If = 0.343	& = 0.003 If = 0.015
	+	& = 0.027 If = 0.172	& = 0.068 If = 0.403	& = 0.027 If = 0.169	& = 0.065 If = 0.393	& = 0.024 If = 0.129	& = 0.063 If = 0.348

For instance, a pessimistic scenario can occur when managerial commitment and PM implementation have simultaneously low levels, which has a simultaneous probability of 0.090, representing a risk for the maintenance manager. However, the probability of the second latent variable occurring since the first variable has happened is 0.569; in other words, if there is a low managerial commitment to implementation of the maintenance program, the probability is 0.569 that the policies focused on machinery and equipment preservation are also at their low level. In order to avoid these scenarios during the TPM implementation process, it is important to have managerial commitment at its high level, since the probability that this scenario is presented along with a low level in PM implementation is only 0.008, which indicates that it will almost never happen, which shows the importance of that variable. Also, the probability of having a low level in PM implementation because there is a high managerial commitment level is 0.048, a very low value that indicates that the second variable in high levels is not associated with low levels in first variable.

The above statement is easily demonstrated when the scenario is analyzed with the variables inverted, that is, when managerial commitment is low and PM implementation is high, which can occur simultaneously at a probability of 0.008, and this indicates that it will almost never happen. Also, it is observed that it is very unlikely that the second variable will occur in its scenario because the first variable has happened, since the probability is only 0.052. However, when managerial commitment and PM implementation have high levels simultaneously, there is a probability of occurrence of 0.084, but the probability that the second variable occurs at its high level since the first variable occurred at its high level is 0.500. The previous data clearly indicate that high managerial commitment levels are related to high PM implementation levels.

How does TPM implementation impact productivity benefits? In order to answer this question, the situation where the first variable is independent and the second is dependent is analyzed. In this case, in the pessimistic environment, when the two variables are at their low level, it is observed that there is a probability of 0.076 that this scenario will occur. However, the probability of having low productivity benefits because the TPM implementation at its low level is 0.475, which indicates that the low levels of the second variable are related to the low levels of the first. Additionally, in an optimistic environment, when TPM implementation and productivity benefits have high levels, the probability that they will occur simultaneously is 0.065, which represents a low value for a maintenance manager; however, the probability that the second variable will occur because the second variable has occurred in its scenario is 0.393. The previous data indicate that TPM implementation is a program that guarantees productivity benefits.

In addition, the previous statement is easily validated when the probability of TPM implementation is at a high level and productivity benefits is at a low level, which has a value of 0.003, indicating that this scenario will almost never occur. Also, the probability that the second variable is at its low level because the first variable is at its high level is only 0.016, which indicates that TPM implementation at a high level is not associated with productivity benefits at a low level. Due to space problems, the interpretation of other scenarios between the latent variables is left to the reader, with an explanation similar to the one that is presented.

A structural equation model was executed, integrating data from the automotive sector, with 172 cases, and another model integrates 159 cases from the aeronautics, electric, electronics, medical, and other sectors. Table 11 shows the β values for every model in the multigroup analysis, for example, for the relationship between managerial commitment and PM implementation for the automotive model it is 0.654 and for the same relationship in the other sectors model it is 0.631, a similar value that needs to be tested statistically for its difference.

Table 11. Beta values for models.

Dependent Variable	Automotive Sector			Other Sectors		
	Independent Variable					
	MC	PMI	TPMI	MC	PMI	TPMI
PM implementation	0.654	-	-	0.631	-	-
TPM implementation	0.49	0.364	-	0.573	0.298	-
productivity benefits	0.277	0.233	0.436	0.304	0.210	0.359

MC, managerial commitment; PMI, PM implementation; TPMI, TPM implementation.

Table 12 shows the confidence intervals for differences between two β values in the analyzed models (automotive sector and other sectors) at a 95% confidence level. For every β value, it is observed that the lower confidence value limit is negative and the upper confidence value limit is positive, and this lets us conclude that there are no differences between analyzed groups, because the zero value is included in that interval.

Table 12. Confidence intervals for differences in β.

Dependent Variable	Independent Variable		
	Managerial Commitment	TPM Implementation	PM Implementation
TPM implementation	−0.098 to 0.1333	-	-
PM implementation	−0.069 to 0.141	−0.096 to 0.115	-
productivity benefits	−0.032 to 0.178	−0.035 to 0.174	−0.104 to 0.107

6. Conclusions and Industrial Implications

Based on the findings previously discussed, the research conclusions are as follows:

1. Based on the R^2 values, TPM implementation has a 62.8% dependence on two variables, but managerial commitment explains most of the variability in 40%. In this sense, manufacturing companies must encourage department leaders and managers to embrace their responsibility for and commitment to TPM. Similarly, managers must promote the active participation of maintenance staff and communicate a corporate vision centered on quality and equipment maintenance, and among these aspects, they must be actively involved in TPM projects. Two other responsibilities of senior managers are to make sure that staff commitment to TPM is aligned with the corporate mission and supervise tracking of the implemented maintenance plans.

2. The managerial commitment latent variable explains 40.7% of PM implementation variability. Therefore, for a preventive maintenance program to be successful, managerial commitment is necessary. Hence, PM programs must be focused on adjusting and changing components before the equipment fails. Also, preventing machine failures must be promoted by managers, since they need to understand the components' life cycle and generate a replacement plan.

3. TPM is a set of programs, among which is preventive maintenance. According to the findings, PM implementation is an important antecedent to any comprehensive TPM program. In fact, in this research, PM implementation explains 22.8% of the variability of TPM implementation. These findings imply that TPM managers and operators must focus their efforts on preventive maintenance programs that consider the components' life cycle to make changes before machines fail.

4. Statistically, three latent variables explain 38.3% of the productivity benefits latent variable: TPM implementation (16.4%), managerial commitment (16.2%), and PM implementation (5.7%). Such estimates imply that managers must pay close attention to the first two variables, since they have the largest effects. Although the direct impact from PM implementation is low in productivity benefits, the indirect effect has a value of 0.097, which can explain 4.55%. In the end, the total effects of PM implementation on productivity benefits have a value of 0.218 units, and this latent variable explains up to 10.2%. In other words, preventive maintenance as a part of TPM implementation is vital if companies aim to obtain productivity benefits.

5. The total effects of managerial commitment are larger than 0.5 standard deviations, demonstrating that this variable is a key element in productivity, TPM success, and PM programs. Consequently, TPM operators must always demand managerial support before starting any preventive maintenance program, because managerial commitment on its own does not guarantee all the productivity benefits, since its direct effects on this variable were only 0.289. On the other hand, the total effects of managerial commitment on productivity benefits where the TPM and PM were involved had a value of 0.582.

6. It is interesting to observe the relationship between managerial commitment and productivity benefits obtained from TPM, where the direct effect was only 0.289, but the indirect effect that occurs through the mediating variables PM implementation and TPM implementation was 0.293, that is, the indirect effect is greater than the direct effect, and the sum gives a total effect of 0.582. The foregoing indicates that management commitment is not sufficient to obtain productivity benefits, because it is necessary to have a labor culture focused on conserving the machinery and equipment that can be reflected in a preventive maintenance program, but in addition, a more holistic TPM implementation program in which all departments of the company are integrated is required.

7. Based on information in Table 10, the following conclusions can be summarized:

    a. High managerial commitment levels are not associated with low productivity benefits levels, even if the probability of simultaneous occurrence is zero.

    b. Even if managerial commitment is low, it is possible to obtain high productivity benefits, because these may come from other sources.

    c. Having low managerial commitment levels represent a risk in PM implementation, TPM implementation, and productivity benefits.

    d. High TPM implementation levels guarantee high productivity benefits levels.

8. There is no statistical evidence to declare that the automotive industrial sector is different from other sectors when multiple groups are analyzed.

7. Research Limitations and Suggestions for Future Work

TPM offers a broad range of benefits for companies, yet this research only analyzes the impact of TPM on productivity. This is one of the limitations in the study, and in order to address it, further research would have to expand the scope by considering other benefits, such as employee safety benefits and organizational benefits. Another limitation in the proposed model and its hypotheses is that it was evaluated using information from the Mexican maquiladora industry, and the same model using data from other geographical areas and industrial sectors may have different results.

Similarly, a successful TPM implementation is not the result of just managerial commitment and PM programs; it is important to explore the impact of other critical success factors, such as machinery technology, employee commitment, employee education and training, tool maintenance, and equipment, and as a result, R-squared is not equal to 1, because another latent variable is not integrated into the model.

Author Contributions: J.R.D.-R. and J.L.G.-A. performed the data collection and data analysis and wrote the paper. J.C.S.D.-M., J.B.-F., and L.A.-S. contributed to the univariate and multivariate analyses and to improving the writing and readability of the paper. Finally, J.R.M.-F. reviewed the English translation and rewrote the paper.

Funding: This project was financed by the Mexican National Council for Science and Technology (CONACYT) under the Thematic Network of Industrial Processes Optimization by grant 330-18-08.

Acknowledgments: The authors acknowledge the maquiladoras companies, managers, and participants who answered the questionnaire in this research.

Conflicts of Interest: The authors declare no conflict of interest.

References

1. Tang, Y.; Liu, Q.; Jing, J.; Yang, Y.; Zou, Z. A framework for identification of maintenance significant items in reliability centered maintenance. *Energy* **2017**, *118*, 1295–1303. [CrossRef]
2. Singh, R.; Gohil, A.M.; Shah, D.B.; Desai, S. Total productive maintenance (TPM) implementation in a machine shop: A case study. *Procedia Eng.* **2013**, *51*, 592–599. [CrossRef]
3. Andersson, R.; Manfredsson, P.; Lantz, B. Total productive maintenance in support processes: An enabler for operation excellence. *Total Qual. Manag. Bus. Excell.* **2015**, *26*, 1042–1055. [CrossRef]
4. Pettersen, J. Defining lean production: Some conceptual and practical issues. *TQM J.* **2009**, *21*, 127–142. [CrossRef]
5. Nakajima, S. *Total Productive Maintenance*; Productivity Press: London, UK, 1998.
6. Attri, R.; Grover, S.; Dev, N.; Kumar, D. An ISM approach for modelling the enablers in the implementation of total productive maintenance (TPM). *Int. J. Syst. Assur. Eng. Manag.* **2013**, *4*, 313–326. [CrossRef]
7. Eti, M.C.; Ogaji, S.O.T.; Probert, S.D. Implementing total productive maintenance in Nigerian manufacturing industries. *Appl. Energy* **2004**, *79*, 385–401. [CrossRef]
8. Rahman, C.M.L. Assessment of Total Productive Maintenance Implementation in a Semiautomated Manufacturing Company Through Downtime and Mean Downtime Analysis. In Proceedings of the 2015 International Conference on Industrial Engineering and Operations Management (IEOM), Aveiro, Portugal, 3–5 March 2015; pp. 1–9.
9. Ahuja, I.P.S.; Khamba, J.S. Total productive maintenance: Literature review and directions. *Int. J. Qual. Reliab. Manag.* **2008**, *25*, 709–756. [CrossRef]
10. Bon, A.T.; Lim, M. Total Productive Maintenance in Automotive Industry: Issues and Effectiveness. In Proceedings of the 2015 International Conference on Industrial Engineering and Operations Management (IEOM), Aveiro, Portugal, 3–5 March 2015; pp. 1–6.
11. Barrios, L.J.; Minguillón, J.; Perales, F.J.; Ron-Angevin, R.; Solé-Casals, J.; Mañanas, M.A. Estado del arte en neurotecnologías para la asistencia y la rehabilitación en españa: Tecnologías auxiliares, trasferencia tecnológica y aplicación clínica. *Revista Iberoamericana de Automática e Informática Industrial RIAI* **2017**, *14*, 355–361. [CrossRef]

12. Ng, K.C.; Goh, G.G.G.; Eze, U.C. Critical Success Factors of Total Productive Maintenance Implementation: A Review. In Proceedings of the 2011 IEEE International Conference on Industrial Engineering and Engineering Management, Changchun, China, 6–9 December 2011; pp. 269–273.

13. Gupta, P.; Vardhan, S.; Haque, M.S.A. Study of Success Factors of TPM Implementation in Indian Industry Towards Operational Excellence: An Overview. In Proceedings of the 2015 International Conference on Industrial Engineering and Operations Management (IEOM), Aveiro, Portugal, 3–5 March 2015.

14. Gómez, A.H.; Toledo, C.E.; Prado, J.M.L.; Morales, S.N. Factores críticos de éxito para el despliegue del mantenimiento productivo total en plantas de la industria maquiladora para la exportación en ciudad juárez: Una solución factorial. *Contaduría y Administración* **2015**, *60*, 82–106. [CrossRef]

15. Shen, C.C. Discussion on key successful factors of TPM in enterprises. *J. Appl. Res. Technol.* **2015**, *13*, 425–427. [CrossRef]

16. Willmott, P.; McCarthy, D. 2—Assessing the true costs and benefits of TPM. In *Total Productivity Maintenance*, 2nd ed.; Butterworth-Heinemann: Oxford, UK, 2001; pp. 17–22.

17. Park, K.S.; Han, S.W. TPM—total productive maintenance: impact on competitiveness and a framework for successful implementation. *Hum. Factors Ergon. Manuf. Serv. Ind.* **2001**, *11*, 321–338. [CrossRef]

18. Bamber, C.J.; Sharp, J.M.; Hides, M.T. Factors affecting successful implementation of total productive maintenance: A UK manufacturing case study perspective. *J. Qual. Maint. Eng.* **1999**, *5*, 162–181. [CrossRef]

19. Netland, T.H. Critical success factors for implementing lean production: The effect of contingencies. *Int. J. Prod. Res.* **2016**, *54*, 2433–2448. [CrossRef]

20. McAdam, R.; Duffner, A.-M. Implementation of total productive maintenance in support of an established total quality programme. *Total. Qual. Manag.* **1996**, *7*, 613–630. [CrossRef]

21. Kamath, N.H.; Rodrigues, L.L.R. Simultaneous consideration of TQM and TPM influence on production performance: A case study on multicolor offset machine using SD model. *Perspect. Sci.* **2016**, *8*, 16–18. [CrossRef]

22. Chlebus, E.; Helman, J.; Olejarczyk, M.; Rosienkiewicz, M. A new approach on implementing TPM in a mine—A case study. *Arch. Civ. Mech. Eng.* **2015**, *15*, 873–884. [CrossRef]

23. Ahuja, I.P.S.; Khamba, J.S. Strategies and success factors for overcoming challenges in TPM implementation in Indian manufacturing industry. *J. Qual. Maint. Eng.* **2008**, *14*, 123–147. [CrossRef]

24. AMAC. Infogramas. Available online: https://indexjuarez.com/estadisticas/infograma (accessed on 12 May 2018). (In Spanish)

25. Caralli, R.; Stevens, J.; Willke, B.; Wilson, W. The Critical Success Factor Method: Establishing a Foundation for Enterprise Security Management. Available online: https://resources.sei.cmu.edu/library/asset-view.cfm?assetid=7129 (accessed on 18 April 2018).

26. Piechnicki, A.S.; Herrero Sola, A.V.; Trojan, F. Decision-making towards achieving world-class total productive maintenance. *Int. J. Oper. Prod. Manag.* **2015**, *35*, 1594–1621. [CrossRef]

27. Hernández Gómez, J.A.; Noriega Morales, S.; Pérez, L.R.; Romero López, R.; Guillen Anaya, L.G. Factores críticos de éxito para la implementación estratégica del MPT: Una revisión de literatura *Ingeniería Industrial. Actualidad y Nuevas Tendencias* **2014**, *4*, 14.

28. Moghaddam, K.S. Preventive maintenance and replacement optimization on CNC machine using multiobjective evolutionary algorithms. *Int. J. Adv. Manuf. Technol.* **2015**, *76*, 2131–2146. [CrossRef]

29. Swanson, L. Linking maintenance strategies to performance. *Int. J. Prod. Econ.* **2001**, *70*, 237–244. [CrossRef]

30. McKone, K.E.; Schroeder, R.G.; Cua, K.O. The impact of total productive maintenance practices on manufacturing performance. *J. Oper. Manag.* **2001**, *19*, 39–58. [CrossRef]

31. Nourelfath, M.; Nahas, N.; Ben-Daya, M. Integrated preventive maintenance and production decisions for imperfect processes. *Reliab. Eng. Syst. Saf.* **2016**, *148*, 21–31. [CrossRef]

32. Eti, M.C.; Ogaji, S.O.T.; Probert, S.D. Reducing the cost of preventive maintenance (PM) through adopting a proactive reliability-focused culture. *Appl. Energy* **2006**, *83*, 1235–1248. [CrossRef]

33. Anis, C.; Nidhal, R.; Mehdi, R. Simultaneous determination of production lot size and preventive maintenance schedule for unreliable production system. *J. Qual. Maint. Eng.* **2008**, *14*, 161–176.

34. Shrivastava, D.; Kulkarni, M.S.; Vrat, P. Integrated design of preventive maintenance and quality control policy parameters with CUSUM chart. *Int. J. Adv. Manuf. Technol.* **2016**, *82*, 2101–2112. [CrossRef]

35. Fumagalli, L.; Macchi, M.; Giacomin, A. Orchestration of preventive maintenance interventions. In Proceedings of the 20th IFAC World Congress, Toulouse, France, 9–14 July 2017; pp. 13976–13981.

36. Ma, L.; Dong, S.; Gong, Y.; Yu, G. Study on application of TPM in small and medium-sized enterprises. In Proceedings of the MSIE 2011, Harbin, China, 6–11 January 2011; pp. 678–681.

37. Ng, K.C.; Chong, K.E.; Goh, G.G.G. Total Productive Maintenance Strategy in a Semiconductor Manufacturer: A Case Study. In Proceedings of the 2013 IEEE International Conference on Industrial Engineering and Engineering Management, Bangkok, Thailand, 10–13 December 2013; pp. 1184–1188.

38. Rodrigues, M.; Hatakeyama, K. Analysis of the fall of TPM in companies. *J. Mater. Process. Technol.* **2006**, *179*, 276–279. [CrossRef]

39. Mwanza, B.G.; Mbohwa, C. Design of a total productive maintenance model for effective implementation: Case study of a chemical manufacturing company. *Procedia Manuf.* **2015**, *4*, 461–470. [CrossRef]

40. Tsuchiya, S. *Quality Maintenance: Zero Defects through Equipment Management*; Productivity Press: London, UK, 1992.

41. Chan, F.T.S.; Lau, H.C.W.; Ip, R.W.L.; Chan, H.K.; Kong, S. Implementation of total productive maintenance: A case study. *Int. J. Prod. Econ.* **2005**, *95*, 71–94. [CrossRef]

42. Smith, R.; Mobley, R.K. Chapter 7—Total productive maintenance. In *Rules of Thumb for Maintenance and Reliability Engineers*; Butterworth-Heinemann: Burlington, VT, USA, 2008; pp. 107–120.

43. Campuzano-Cervantes, J.; Meléndez-Pertuz, F.; Núñez-Perez, B.; Simancas-García, J. Sistema de monitoreo electrónico de desplazamiento de tubos de extensión para junta expansiva. *Revista Iberoamericana de Automática e Informática Industrial RIAI* **2017**, *14*, 268–278. [CrossRef]

44. Abhishek, J.; Rajbir, B.; Harwinder, S. Total productive maintenance (TPM) implementation practice: A literature review and directions. *Int. J. Lean Six Sigma* **2014**, *5*, 293–323.

45. Nakajima, S. *Introduction to TPM: Total Productive Maintenance*; Productivity Press: Cambridge, MA, USA, 1988.

46. Nakajima, S. *TPM Development Program: Implementing Total Productive Maintenance*; Productivity Press: Cambridge, MA, USA, 1989.

47. Lycke, L. Team development when implementing TPM. *Total. Qual. Manag. Bus. Excell.* **2003**, *14*, 205–213. [CrossRef]

48. Ahmad, N.; Hossen, J.; Ali, S.M. Improvement of overall equipment efficiency of ring frame through total productive maintenance: A textile case. *Int. J. Adv. Manuf. Technol.* **2018**, *94*, 239–256. [CrossRef]

49. Chong, M.Y.; Chin, J.F.; Hamzah, H.S. Transfer of total productive maintenance practice to supply chain. *Total Qual. Manag. Bus. Excell.* **2012**, *23*, 467–488. [CrossRef]

50. Ahuja, I.P.S.; Khamba, J.S. An evaluation of TPM initiatives in Indian industry for enhanced manufacturing performance. *Int. J. Qual. Reliab. Manag.* **2008**, *25*, 147–172. [CrossRef]

51. Kalir, A.A.; Rozen, K.; Morrison, J.R. Evaluation of preventive maintenance segregation: A multi factorial study. *IEEE Trans. Semicond. Manuf.* **2017**, *30*, 508–514. [CrossRef]

52. Chalabi, N.; Dahane, M.; Beldjilali, B.; Neki, A. Optimisation of preventive maintenance grouping strategy for multi-component series systems: Particle swarm based approach. *Comput. Ind. Eng.* **2016**, *102*, 440–451. [CrossRef]

53. Avelar-Sosa, L.; García-Alcaraz, J.L.; Vergara-Villegas, O.O.; Maldonado-Macías, A.A.; Alor-Hernández, G. Impact of traditional and international logistic policies in supply chain performance. *Int. J. Adv. Manuf. Technol.* **2015**, *76*, 913–925. [CrossRef]

54. Duenckel, J.R.; Soileau, R.; Pittman, J.D. Preventive maintenance for electrical reliability: A proposed metric using mean time between failures plus finds. *IEEE Ind. Appl. Mag.* **2017**, *23*, 45–56. [CrossRef]

55. Leys, C.; Ley, C.; Klein, O.; Bernard, P.; Licata, L. Detecting outliers: Do not use standard deviation around the mean, use absolute deviation around the median. *J. Exp. Soc. Psychol.* **2013**, *49*, 764–766. [CrossRef]

56. Hair, J.F.J.; Black, W.C.; Babin, B.J.; Anderson, R.E. *Multivariate Data Analysis*; Prentice Hall: Upper Saddle River, NJ, USA, 2013.

57. Lynch, S.M. *Introduction to Applied Bayesian Statistics and Estimation for Social Scientists*; Springer Science & Business Media: New York, NY, USA, 2007.

58. Kohler, M.; Müller, F.; Walk, H. Estimation of a regression function corresponding to latent variables. *J. Stat. Plan. Inference* **2015**, *162*, 22. [CrossRef]

59. Tabachnick, B.; Fidell, L. *Using Multivariate Statistics*, 2nd ed.; Pearson: Chennai, India, 2013.

60. Kock, N. *Warppls 5.0 User Manual*; ScriptWarp Systems: Laredo, TX, USA, 2015.

61. Adamson, K.A.; Prion, S. Reliability: Measuring internal consistency using Cronbach's α. *Clin. Simul. Nurs.* **2013**, *9*, e179–e180. [CrossRef]

62. Evermann, J.; Tate, M. Assessing the predictive performance of structural equation model estimators. *J. Bus. Res.* **2016**, *69*, 4565–4582. [CrossRef]

63. Jenatabadi, H.S.; Ismail, N.A. Application of structural equation modelling for estimating airline performance. *J. Air Transp. Manag.* **2014**, *40*, 25–33. [CrossRef]

64. Kock, N. Advanced mediating effects tests, multi-group analyses, and measurement model assessments in PLS-based SEM. *Int. J. e-Collab.* **2014**, *10*, 1–13. [CrossRef]

65. Midiala Oropesa, V.; Jorge Luis García, A.; Aidé Aracely Maldonado, M.; Valeria Martínez, L. The impact of managerial commitment and kaizen benefits on companies. *J. Manuf. Technol. Manag.* **2016**, *27*, 692–712. [CrossRef]

66. García-Alcaraz, J.L.; Prieto-Luevano, D.J.; Maldonado-Macías, A.A.; Blanco-Fernández, J.; Jiménez-Macías, E.; Moreno-Jiménez, J.M. Structural equation modeling to identify the human resource value in the JIT implementation: Case maquiladora sector. *Int. J. Adv. Manuf. Technol.* **2015**, *77*, 1483–1497. [CrossRef]

67. Mendoza-Fong, J.R.; García-Alcaraz, J.L.; Ochoa-Domínguez, H.d.J.; Cortes-Robles, G. Green production attributes and its impact in company's sustainability. In *New Perspectives on Applied Industrial Tools and Techniques*; García-Alcaraz, J.L., Alor-Hernández, G., Maldonado-Macías, A.A., Sánchez-Ramírez, C., Eds.; Springer: Cham, Switerland, 2018; pp. 23–46.

applied
sciences

MDPI

Article

Opportunities for Industry 4.0 to Support Remanufacturing

Shanshan Yang, Aravind Raghavendra M. R. *, Jacek Kaminski and Helene Pepin

Advanced Remanufacturing and Technology Centre (ARTC), Agency for Science, Technology and Research (A*STAR), 3 CleanTech Loop, #01/01, CleanTech Two, Singapore 637143, Singapore; yangs@artc.a-star.edu.sg (S.Y.); kaminskijk@artc.a-star.edu.sg (J.K.); helene-pepin@artc.a-star.edu.sg (H.P.)
* Correspondence: aravindr@artc.a-star.edu.sg; Tel.: +(65)-6908-7911

Received: 28 June 2018; Accepted: 13 July 2018; Published: 19 July 2018

Abstract: Remanufacturing is the process of bringing end-of-life products back to good-as-new. It plays a critical role in decoupling economic growth from growth in resource use, and in accelerating the circular economy. However, the uptake of remanufacturing activities faces obstacles. This paper reviews the challenges encountered by the remanufacturing sector and discusses how the Industry 4.0 revolution could help to effectively address these issues and unlock the potential of remanufacturing. Two case studies are included in this paper to exemplify how technology enablers from Industry 4.0 can increase efficiency, reliability, and digitization of the remanufacturing process.

Keywords: remanufacturing; sustainable manufacturing; Industry 4.0; smart factory; intelligent machining

1. Introduction

The growth and stability of the traditional "take-make-consume-dispose" linear model has been heavily dependent on the availability of resources. However, the linear model is now challenged by an unprecedented rise in demand for the finite supply of resources, as it is expected that by 2030 there will be three billion more middle-class consumers worldwide [1]. With this in mind, the "circular economy" model is drawing global attention as an approach to decouple economic growth from resource constraints. The underlying principle of the "circular economy" is to make products and materials restorative and regenerative by design, and to maintain them at their highest utility and value [2].

Circulation of the technical product life cycle is enabled through several end-of-life (EOL) channels, including recycling, remanufacturing, reusing, refurbishing, etc. Among these EOL strategies, remanufacturing shows noticeable advantages due to its effectiveness in preserving the added value of products and assuring their quality. Remanufacturing is the process of bringing EOL products back to good-as-new status through disassembly, cleaning, inspection/sorting, restoring, and reassembly. It provides social, environmental, and economic benefits by offering to consumers end products with assured quality at a competitive price, while protecting the intellectual property and brand image of original equipment manufacturers (OEMs). It also creates and opens up new business and job opportunities in the after-sales service market. Another main advantage of remanufacturing lies in protection of the environment by reducing the usage of raw materials, carbon footprint, and number of components being scrapped. However, the uptake of remanufacturing faces several obstacles, which need to be resolved properly through collaboration among multiple players across the business, government, investors, society, and research communities [3]. Fortunately, the advent of Industry 4.0 has provided immense opportunities for unlocking the potential for remanufacturing by reducing the cost of transformation into a higher level of connectivity and efficiency.

Industry 4.0, as the name suggests, refers to the fourth stage of industrialization, aiming for a high level of automation in the manufacturing industry through the adoption of ubiquitous information and communication technologies (ICTs). The boundaries between the virtual environment and real world get increasingly blurred, which is called "cyber-physical production systems (CPPSs)". In simple terms, in CPPSs, electronic and mechanical components are linked through sensors in a network, which provides a smart platform for data flow and data analytics. An early form of this technology is the implementation of radio frequency identification (RFID) sensors, which have been in wide use since the year 1999.

In this paper, the opportunities that Industry 4.0 bring to the remanufacturing industry are discussed and presented from the perspectives of "Smart Life Cycle Data", "Smart Factory", and "Smart Services". Case studies are presented to exemplify the use of ICT to increase the efficiency and accuracy of the remanufacturing process.

2. Literature Review

Remanufacturing activities span a number of industries. Remanufacturing has experienced a higher chance for success in sectors in which:

- Products are durable and usually contain high-value materials;
- The technology cycle is stable and longer than the useful life cycle;
- Restoration technologies are available;
- Products have the potential to be leased or delivered as a service rather than as hardware.

The above-mentioned factors are also the reasons why more than 60% of remanufacturing activities, in terms of production value, are concentrated in the aerospace, automotive, and heavy-duty off-road vehicle sectors. Examples of products or components being remanufactured include aerospace engines, alternators for automotive, gears for heavy-duty vehicles, etc. More recently, driven by noticeable economic profits and ever more stringent environmental regulations, remanufacturing has also been carried out by other industries, such as IT products, machinery, tires, and furniture.

Bottlenecks to remanufacturing vary by geographic location, industry sectors size of the individual remanufacturer, and nature of business. Nevertheless, there are still some common challenges, which are identified as follows:

- Lack of standards and legislation

The lack of a commonly accepted definition and standards for remanufactured products in various sectors has been identified as the most prevalent barrier to remanufacturing. This has resulted in not only the loss of consumer trust in remanufactured products, but also restrictions on importing or exporting remanufactured products in certain countries [4].

- Lack of life cycle design awareness

Many barriers encountered during the remanufacturing process could be eliminated if proper design features were included in the early stage of product design. For example, avoiding the use of permanent joints reduces the complexity of disassembly processes, and standardizing part designs simplifies the inspection and sorting process. Currently, most design for remanufacturing (DfRem) tools or concepts remain within the realm of academic research. There is still a lack of an effective DfRem tool commonly accepted and adopted by industry, which hinders the closing of the product life cycle through remanufacturing [5].

- Lack of sufficient market demand and core supply

"Cores" refer to the used products which are re-collected for remanufacturing. A lack of understanding and negative perception of remanufacturing have limited market demand for remanufactured products. When customers are not convinced of the quality of remanufactured products, they expect remanufactured products to be sold at a lower price, which consequently reduces the profit margin and discourages growth of the remanufacturing industry. Furthermore, due to the current linear movement of product from the point of manufacture to customers and, ultimately, to landfills, incineration, or material extraction, gaining access to EOL products and diverting qualified ones to a remanufacturing facility also pose challenges to the growth of remanufacturing. There is a need to change to circular or service-oriented models that can truly reverse the linear trend and assure a supply of cores [4].

- Skill/technology challenges and limited information sharing

Remanufacturing is a highly labor-intensive industry in comparison with traditional manufacturing [6]. Many of the decisions made during the remanufacturing process are still dependent on ad hoc engineering experience and thus require technically skilled engineers or technicians. Meanwhile, there is still a need for the development of suitable nondestructive testing (NDT) methods for repaired components or parts, which makes the qualification of remanufactured products challenging. In addition to this, the lack of original design specifications and information on the usage and repair history of the returned products further complicate the assessment of the viability of core remanufacturing.

Even though progress towards the uptake of remanufacturing is still hampered by a few challenges, Industry 4.0 reveals some powerful emerging trends that have the potential to address these bottlenecks. While Industry 3.0 focused on the automation of a single machine and process, Industry 4.0 pushes towards end-to-end digitalization of all physical assets and the entire supply chain [7]. The rise of Industry 4.0 is leading to the drastic and rapid growth of data volume (Big Data), driven by:

- The availability of computing power and connectivity,
- The advancement of analytic capability (e.g., artificial intelligence),
- The introduction of new patterns of human and machine interaction (e.g., augmented reality systems), and
- The advent of technologies that ease the transformation of digital data into physical objects (e.g., additive manufacturing and rapid prototyping).

When it comes to digitalization, Industry 4.0 possesses four main characteristics:

- The vertical networking of small production systems, such as smart factories and smart products;
- The horizontal integration of global value creation networks, such as new business and cooperation models;
- Through-engineering across the product design, product use, and EOL stages;
- Exponential acceleration of technologies.

The fundamental idea of Industry 4.0 is to increase availability and integrated use of relevant data by connecting all products, resources, and companies involved in the value chain and, ultimately, to generate additional value from available data and to maximize customer benefit [8]. According to a recent global industry survey [7], Industry 4.0 is no longer considered a "future trend" and for many companies it is already the core of their strategy and research agenda. Around 40% of companies surveyed reported that their vertical and horizontal value chains and through-engineering model are already benefiting from an advanced level of digitization and integration; this number is expected to increase to 75% within 5 years. Meanwhile, survey results suggested that global industrial product companies would invest USD907 billion per year through to 2020 for digital transformation. It is believed that the digitization of horizontal and vertical value chains, as well as the product life cycle, will eventually revolutionize the product, manufacturing/remanufacturing industry, and corresponding business models.

3. Smart Remanufacturing in the Digital Age

To pave the way towards a wider uptake of remanufacturing, the challenges discussed in the previous section need to be addressed and properly resolved by OEM and independent remanufacturers. Fortunately, the advent of Industry 4.0 has presented immense opportunities to unlock the potential of remanufacturing. In this section, the opportunities that Industry 4.0 bring to remanufacturing are discussed based on the three aspects previously mentioned, namely, Smart Life Cycle Data for Design for Remanufacturing and EOL Management, Smart Factory for cost-effective and green remanufacturing operations, and Smart Services for a successful remanufacturing business model. Figure 1 describes the three application areas and technical enablers from Industry 4.0. The technology enablers are presented in the outer rim of the circle, and include smart sensors, cloud computing, robotics, machine-to-machine communication (M2M), additive manufacturing, and others.

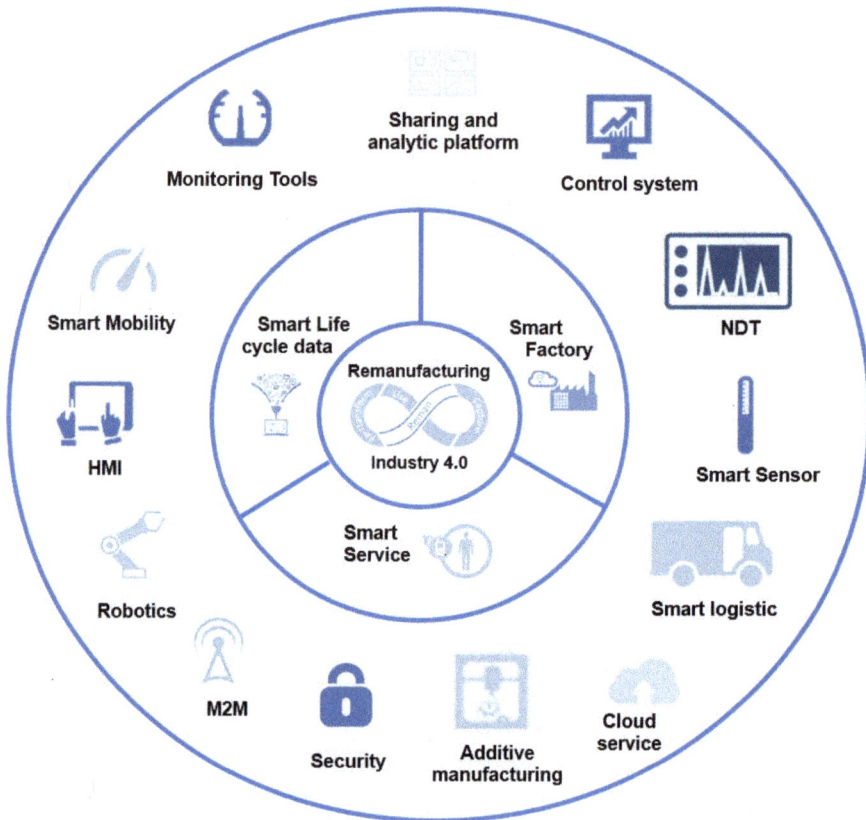

Figure 1. Opportunities from Industry 4.0 for remanufacturing, and its key enablers.

3.1. Smart Life Cycle Data

From the initial product design and development stage, and all the way until the EOL stage, various product information is generated and captured. Ideally, this information should be shared across the product life-cycle stakeholders to support product life-cycle management [5]. However, the flow of product information remains essentially unestablished due to ineffective data extraction, loss of information during product transfer between stakeholders, undeveloped platforms to support information sharing, and other policy restrictions. Ineffectiveness of data circulation has reduced the efficiency of product life-cycle management and the quality of service provided. For example, incomplete information on returned cores remains a big challenge for the majority of remanufacturers [5]. In order to restore these products back to "good as new" quality, remanufacturers have to recreate product knowledge which existed at the product design stage. In this regard, the digital transformation of Industry 4.0 has shed some light on addressing this concern by improving data transferability and building the knowledge/data-sharing platform. This could be enabled through sensors, embedded systems, and connected devices ("Internet of Things"), as well as a comprehensive data management platform. For example, when information regarding computer-aided design (CAD), bill of materials, parts information, manufacturing and assembly instructions, data from product use stage, and repair history information are stored in a central system and are easily accessible by remanufacturers, the repair decisions during the remanufacturing process could be made easily and operations could be carried out in a more efficient manner. Furthermore, when information, such as product failure modes and rates, replacement frequency, cleaning efficiency, disassembly challenges, and upgrading challenges, is extracted effectively from the remanufacturing stage and fed back to product designers, many of the barriers occurring during the remanufacturing process could be avoided in the next generation of products by incorporating proper design features [9–12]. This is strategically important and a substantial cost-saving measure, as more than 70% of product costs are determined at the product development stage [13].

3.2. Smart Factory

Due to the uncertainty involved in the number and quality of cores returned, remanufacturing operations need to incorporate a high degree of flexibility to react quickly and appropriately to various product reconditioning requirements. Smart factories, which enable high flexibility and small batch size production, seamlessly address this complexity issue associated with remanufacturing operations. "Smart factories" are essentially at the core of Industry 4.0. "Smartness" is achieved using electronic hardware/software, as well as networking of production resources. Compared with traditional manufacturing, more ancillary hardware and software, like RFID tags, barcodes, laser markers, sensors, as well as communication infrastructure, will be embedded into the factory to enable machines to collaborate with each other using intelligent analytics. In the future, in a smart remanufacturing environment, machines could obtain incoming core information through scanning a barcode attached to the core, adapt the remanufacturing operations through self-optimization and smart fixturing capabilities, update the process-related information to a database via wireless transfer, and store remanufacturing knowledge gained from experience. This could enable a substantial reduction of the labor force and lessen the dependency on high-skilled operators. In the meanwhile, with various types of sensors embedded into equipment and cells, data from the manufacturing process could be retrieved and sent for real-time analysis. This would support the early detection of machines or cell failures and allow preventative strategies to be implemented to avoid unplanned maintenance and catastrophic failures. Information associated with the product manufacturing data could also been recorded and stored as part of the product life-cycle data. In addition, energy-efficient remanufacturing processes could also be achieved through collecting real-time energy consumption data and implementing an energy management system accordingly. Further innovative technologies, such as additive manufacturing, 3D scanning, automated guided vehicles, inspection drones, hybrid manufacturing/process [14,15], and augmented reality tools, will continuously drive down the cost of remanufacturing operations while also delivering substantial improvements in the quality of the repaired product. Taking aerospace remanufacturing as an illustration, worn airfoils can sometimes be repaired using the laser metal deposition technique, an additive process where metal powder is melted by a computer numerical control (CNC) laser on a robot arm to form a direct fusion-bonded deposit on the blades. This technology offers the precision and low heat input necessary to successfully achieve such restoration, compared to more conventional fusion welding processes. In addition, sustainability needs to be an important factor to be considered while adopting these advanced technologies. The life-cycle assessment method could be conducted to gauge the environmental impact of these new technologies to make sure the advancement is not only technically achieved but is also environmentally friendly [16].

3.3. Smart Services

One of the challenges that remanufacturers face, as explained earlier, is the control of the timing, quality, and quantity of cores returned. In this regard, the product service system has provided opportunities to cope with the complexity of core return for remanufacturing [17]. In this emerging and disruptive business model, ownership of the product is usually retained by the OEMs or retailers and only the service or usage is offered to customers (e.g., selling "flying hours of the engine" instead of selling "engines"). Hence, it creates a mandate for manufacturers or retailers to monitor their product's performance during its runtime and to forecast remanufacturing operations on the cores returned based on the predicted remaining life of the product. On the other hand, from the consumers' perspective, as they pay for the service rather than the ownership of the product, market acceptance for remanufactured goods will likely be increased, leading to a successful remanufacturing model. Real-time monitoring of product in-use and data analysis via embedded sensor networks and cloud-based computation could enable predictive maintenance by the early detection of problems. Increased connectivity among products, customers, and manufacturers, promoted by Industry 4.0, presents immense opportunities for boosting the product service model. Take, for example, machines that are leased to the power plant sector. Power plants rely heavily on the continuous availability of their machinery. To increase the reliability of machines, smart sensors are embedded into the machines to monitor critical factors, such as temperature, pressure, switches, energy consumption, and vibration in real time [18]. Sensor data are collected and logged into a central server through a network for prediction of potential wear and estimation of components' useful life, and for scheduling maintenance or remanufacturing of components in a timely manner.

4. Case Illustrations

As mentioned previously, traditional remanufacturing is hindered by low process efficiency and data silos among various product life-cycle stages. In this section, we present two case studies to demonstrate how Industry 4.0 helps traditional remanufacturing achieve better connectivity between machines through seamless data flow and real-time data capturing for condition monitoring, respectively.

4.1. Smart Repair Cell

Currently, most remanufacturing tasks are highly labor intensive, as returned cores vary greatly in terms of their form and condition. Even though human intelligence is capable of adapting to such variations and performing adaptive manual repair tasks, dealing with variability in the part profile and form accuracy of each core still remains challenging for the remanufacturing process. This is particularly a concern for aerospace part repair, as the sector has stringent requirements for compliance to the original part specifications.

To achieve consistency and high flexibility in remanufacturing operations, a smart remanufacturing cell, with special focus on repair operations, has been simulated at the Advanced Remanufacturing and Technology Centre (ARTC). The smart cell, as graphically described in Figure 2, comprises the following layers:

- *Shared database*—in this layer, all virtual data pertaining to every individual repaired component are stored and organized based on the design and existing repair information. By organizing the repair data, backtracking the core's repair process history and failure information is a well-established process.

- *Software platform*—this layer defines all the software tasks related to the data flow and machine-to-machine communication channels. With a common software platform, high-speed computations are carried out both in real time and offline, helping seamless data transfer between machines without any loss in data quality.

- *Physical process*—this layer of the smart cell defines the actual repair tasks to be carried out on the core. Since the process requires physical transfer of the component and its storage container across the actual repair processes, it relies heavily on factory automation to conform to Industry 4.0 standards.

In standard manufacturing protocols, design of the component using computer-aided design (CAD) precedes part manufacturing. Yet, in the remanufacturing process, the type and level of damage to the core will affect repair decisions. Some of the damage modes include, but are not limited to: cracking, wear, physical distortion, high-temperature damage, surface corrosion, etc. Upon entering the cell, cores will be registered in the shared database using identification markers during a process termed *"Part Registration"*. This process provides the system with core information, such as original part number, serial number, raw material identification, bill of materials (if it is an assembly), OEM CAD design information, and repair history.

Once cores are registered, a single/series of repair processes will be planned for each core in a sequence suitable for its identification and condition, based on established standard OEM repair protocols or other approved methods. The repair sequence will involve the utilization of several machines and manufacturing processes, such as brazing, thermal/cold spray, and laser cladding to regenerate the damaged area of the core. Process-related information will be updated in the database via wireless transfer and will constitute the digital "life-story" of the parts. This registration process is crucial in product life-cycle management (PLM) to improve the product design in the future and to ensure an appropriate decision-making process at the next remanufacturing cycle.

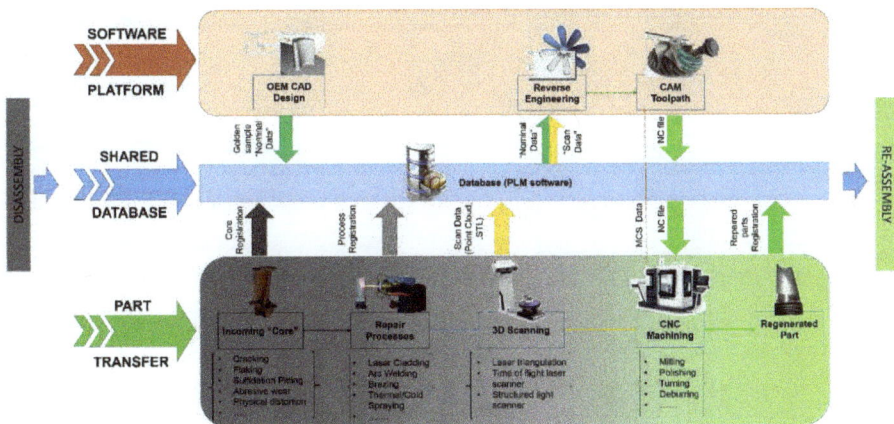

Figure 2. Smart remanufacturing cell-process flow.

As most of the repair processes result in addition of material to the parts, further machining processes need to be carried out to bring the repaired component back to its original dimensions. This is a critical stage of the repair process, as it will determine the accuracy and tolerances of the final repaired part. To generate the optimum tool path, automated adaptive machining is adopted. Firstly, the as-received physical dimensions of the core need to be defined in 3D digital format. This could be realized using various three-dimensional scanning methods, such as laser triangulation, structured light, probing methods, etc., to create the digital twin of the core. The scanned output is stored in the shared database, usually in point-cloud form. Secondly, the scanned data is compared with the original digital design of the part (computer-aided model—CAD), also called golden sample/nominal data, to obtain the geometry difference, using reverse engineering software. Thirdly, on-machine probing or a scanning method is carried out to measure the information on the machine coordinate system (MCS) and to align the repaired part with respect to the machine datum. Lastly, via the developed machine-to-machine communication software platform, geometry differences and the MCS are fed into computer-aided manufacturing (CAM) software for the adaptive tool-path generation. The tool path generated based on this aligned datum is stored in the shared database layer as a numerical control (NC) file. Following the additive repair processes, when a part is ready for final machining, the NC file is retrieved from the database to be used for machining. Various machining processes, such as 5-axis milling, turning, grinding, abrasive polishing, etc., are carried out to achieve the final dimensions on the repaired component. Machining-related information is also stored in the database as part of the product's digital "life-story". The real-time monitoring of the machining data plays a key role in defining the surface integrity of the machined parts. In the subsequent case study, the effect of sensor integration on machine tools and its impact on Industry 4.0 is discussed in detail.

In addition, thanks to the advancement of machine tool technology, both additive and subtractive processes can be performed with one single machine. Hybrid machines—combining both laser metal deposition blown powder and 5-axis milling in one machine—are now commercially available for use in production. This provides the benefit of single-machine and single-work setup, hence improving the volumetric accuracy of the final repaired component. In addition, the cost of multiple machines, corresponding part transfer automation, and the space requirement are reduced substantially.

Smart repair cells are designed to be an effective way of repairing each component adaptively with utmost accuracy and tolerance due to seamless information flow across the processes. Currently, a smart repair cell concept is simulated at ARTC using various scenarios such as number of machine tools, robotic automation vs manual part transfer, and volume of cores. Based on the simulated results, the best operational efficiency is calculated to select the right hardware combinations. Know-how in reverse engineering, adaptive toolpath development, and individual repair processes have already been embedded into the smart repair cell. Ongoing work at ARTC is focused on building up the database based on number of cores, real-time machine tool monitoring, and the final remanufactured part conformance data, which will eventually help machine learning for automatic repair decision making. It should be noted that the above-mentioned remanufacturing cell focuses mainly on the adaptive repair process. Disassembly, cleaning, re-assembly, and, most critically, the NDT process for analyzing issues related to adhesion of the repaired material, change of material structure, porosity, inclusions, etc. are not addressed in this paper.

4.2. Sensorized Machines for Intelligent Machining

Machine tools could become increasingly smart by perceiving their own states and the state of the surrounding environment, which is considered as "intelligent machining". Key enablers for this capability include smart sensors embedded into machine systems, a data acquisition system which can transmit the processing data quickly without data loss, and also the intelligent central system which receives and analyzes the data. Such an intelligent machining system is of great importance for evaluating the system's health and to enable prediction of a possible breakdown or malfunction before it happens. Currently, advanced condition-based maintenance is triggered when an asset reaches a predefined unacceptable level detected by the sensing system. These thresholds usually originate from experts' experiences (machine operators) or manufacturers' recommendations. However, when multiple complicating factors coexist, false alarms might be triggered. To resolve such complications, sensors and advanced computer systems will be used in tandem during intelligent machining to monitor the system health. Both historical and real-time sensor data will be used in the evaluation of the machine condition via artificial intelligence techniques to substantially increase the prediction accuracy and the quality of the produced part. This work is now being carried out by a research group in ARTC.

A large number of sensing devices, including vibration sensors, current sensors, temperature sensors, and acoustic emission sensors, have been installed into various critical subsystems of a computer numerical control (CNC) machine, such as spindles, linear guides, ball screws, cutting tools, etc. All the sensors are linked to three data acquisition (DAQ) systems. Each DAQ system performs data acquisition and storage. The data are used for analyzing the correlation between machine health and the quality of the product selected to be machined. Sensor data are captured and recorded in the database during the process of machining the product, which is, in our case, a carburized steel shaft.

As modern machines are quite robust, it may take months or years before detectable changes in machine operation and performance can be observed. Sensors can register small changes indicating the beginning of degradation in machine components or subsystems. However, it is not guaranteed that such changes will affect the quality of machined products. Therefore, it is difficult to determine the remaining useful life of a whole machine system or subsystem based on one-time measurement without access to historical data. In order to obtain the critical and more relevant machine data, different levels of failure modes, such as tool wear, tailstock misalignment, and spindle unbalance, were simulated. At the same time, changes in sensor data collected during experiments to simulate different levels of machine failure modes, as well as the measurement of part quality, including surface roughness, waviness, dimensions, and out-of-roundness, have been captured and documented in order to search for correlations between the sensor signals, deterioration of machine subsystems, and quality of machined products, as shown in Figure 3. To date, a total of 190 GB of data have been collected during the machining of 45 shafts under the different simulated failure modes. Work is ongoing to analyze these data and to generate a model for machine failure prediction.

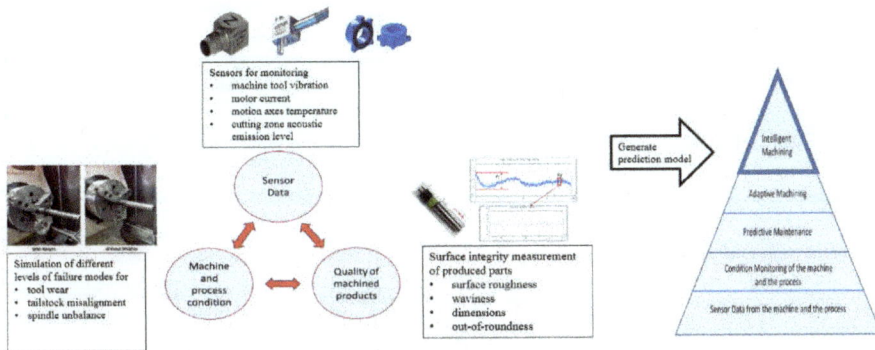

Figure 3. Correlation between machine and process condition, sensor data, and quality of machined products.

The purpose of developing intelligent machining capability is to be able to monitor machine condition and part quality, to predict the need for maintenance, and to avoid costly failure and the need for part rework. In the meanwhile, data captured will also form part of a product's "digital life-story", which will be of great use for supporting product end-of-life decision making and analyzing the cause of product failure. Additionally, the ability to retrofit older machines with smart sensors and the data acquisition system to achieve intelligent machining is of paramount importance, as not all companies can afford to purchase new modern machinery to enjoy the benefits of the sensor system. The capability of retrofitting an existing machine with an advanced sensor system has been developed and demonstrated through this current research work.

5. Conclusions

In this paper, some of the opportunities that Industry 4.0 bring to the remanufacturing industry are discussed and presented from the perspectives of "Smart Life Cycle Data", "Smart Factory", and "Smart Service". It is observed that increased digitalization across the supply chain and enhanced cyber-physical intelligence within the factory have effectively addressed several major concerns that remanufacturers encounter. This will potentially reduce the cost of transformation in remanufacturing.

A smart remanufacturing cell, which provides tailor-made repair operations based on incoming core conditions, is described in the case study section to showcase technology enablers from Industry 4.0 supporting the smart remanufacturing process. Smart sensors which are embedded into the CNC machine enable the real-time monitoring of machine health and call for maintenance or component remanufacturing in a preventative manner. Future work could also look at the economic analysis to justify the viability and profitability of utilizing innovative technologies for a smart remanufacturing system.

Author Contributions: Conceptualization, S.Y., A.R.M.R. and J.K.; Methodology, S.Y.; Supervision, H.P.; Writing-original draft, S.Y., A.R.M.R. and J.K.; Writing-review & editing, S.Y., A.R.M.R., J.K. and H.P.

Funding: This research received no external funding.

Conflicts of Interest: The authors declare no conflict of interest.

References

1. World Economic Forum. Towards the Circular Economy: Accelerating the Scale-up across Global Supply Chains. 2014. Available online: http://reports.weforum.org/toward-the-circular-economy-accelerating-the-scale-up-across-global-supply-chains/ (accessed on 28 June 2018).

2. Circular Economy, Ellen Macarthur Foundation. 2018. Available online: https://www.ellenmacarthurfoundation.org/circular-economy (accessed on 28 June 2018).

3. Parker, D.; Riley, K.; Robinson, S.; Symington, H.; Tewson, J.; Jansson, K.; Ramkumar, S.; Peck, D. Remanufacturing Market Study. 2015. Available online: http://www.remanufacturing.eu/assets/pdfs/remanufacturing-market-study.pdf (accessed on 28 June 2018).

4. APSRG & APMG, Triple Win—The Social, Economic and Environmental Case for Remanufacturing. 2014. Available online: http://www.policyconnect.org.uk/apsrg/sites/site_apsrg/files/triple_win_-the_social_economic_and_environmental_case_for_remanufacturing.pdf (accessed on 28 June 2018).

5. Kurilova-Palisaitiene, J.; Lindkvist, L.; Sundin, E. Towards facilitating circular product life-cycle information flow via remanufacturing. *Procedia CIRP* **2015**, *29*, 780–785. [CrossRef]

6. Statham, S. Remanufacturing Towards a More Sustainable Future. In *Electronics-Enabled Products Knowledge-Transfer Network*; Loughborough University: Loughboroug, UK, 2006.

7. PwC. Industry 4.0: Building the Digital Enterprise. 2015. Available online: https://www.pwc.com/gx/en/industries/industries-4.0/landing-page/industry-4.0-building-your-digital-enterprise-april-2016.pdf (accessed on 28 June 2018).

8. Koch, V.; Kuge, S.; Geissbauer, R.; Schrauf, S. Industry 4.0: Opportunities and Challenges of the Industrial Internet. 2015. Available online: https://www.strategyand.pwc.com/media/file/Industry4.0.pdf (accessed on 28 June 2018).

9. Nasr, N.; Thurston, M. Remanufacturing: A key enabler to sustainable product systems. In Proceedings of the 13th CIRP International Conference on Life-Cycle Engineering, Leuven, Belgium, 31 May–2 June 2006; pp. 15–18.

10. Amezquita, T.; Hammond, R.; Salazar, M.; Bras, B. Characterizing the remanufacturability of engineering systems. In Proceedings of the ASME Advances in Design Automation Conference, Boston, MA, USA, 17–20 September 1995; pp. 271–278.

11. Ijomah, W.; McMahon, C.; Hammond, G.; Newman, S. Development of robust design-for-remanufacturing guidelines to further the aims of sustainable development. *Int. J. Prod. Res.* **2007**, *45*, 4513–4536. [CrossRef]

12. Sundin, E. Product and Process Design for Successful Remanufacturing. Ph.D. Thesis, Linköping University, Linköping, Sweden, 2004.

13. Anderson, D.M. *Design for Manufacturability: How to Use Concurrent Engineering to Rapidly Develop Low-Cost, High-Quality Products for Lean Production*; CRC Press: Boca Raton, FL, USA, 2014.

14. Calleja-Ochoa, A.; Gonzalez-Barrio, H.; Polvorosa-Teijeiro, R.; Ortega-Rodriguez, N.; Lopez-de-Lacalle-Marcaide, L.N. Multitasking machines: evolution, resources, processes and scheduling. *DYNA* **2017**, *92*, 637–642.

15. Urbikain, G.; Perez, J.M.; López de Lacalle, L.N.; Andueza, A. Combination of friction drilling and form tapping processes on dissimilar materials for making nutless joints. *Proc. Inst. Mech. Eng. Part B* **2018**, *232*, 1007–1020. [CrossRef]

16. Pereira, O.; Martín-Alfonso, J.E.; Rodríguez, A.; Calleja, A.; Fernández-Valdivielso, A.; de Lacalle, L.L. Sustainability analysis of lubricant oils for minimum quantity lubrication based on their tribo-rheological performance. *J. Clean. Prod.* **2017**, *164*, 1419–1429. [CrossRef]

17. Guidat, T.; Barquet, A.P.; Widera, H.; Rozenfeld, H.; Seliger, G. Guidelines for the definition of innovative industrial product-service systems (PSS) business models for remanufacturing. *Procedia CIRP* **2014**, *16*, 193–198. [CrossRef]

18. Rese, M.; Everhartz, J. Condition Monitoring of Industrial Product Service Systems–Helpful selling argument or potential marketing pitfall? In *The Philosopher's Stone for Sustainability*; Springer: Berlin/Heidelberg, Germany, 2013; pp. 493–497.

applied
sciences

MDPI

Article

Hybrid Laminate for Haptic Input Device with Integrated Signal Processing

René Schmidt [1,*], Alexander Graf [2], Ricardo Decker [3], Verena Kräusel [2], Wolfram Hardt [1], Dirk Landgrebe [2] and Lothar Kroll [3]

1   Computer Engineering, Chemnitz University of Technology, Straße der Nationen 62, 09111 Chemnitz, Germany; cera@cs.tu-chemnitz.de
2   Forming and Joining, Chemnitz University of Technology, Reichenhainer Straße 70, 09107 Chemnitz, Germany; alexander.graf@mb.tu-chemnitz.de (A.G.); verena.kraeusel@mb.tu-chemnitz.de (V.K.); uff@mb.tu-chemnitz.de (D.L.)
3   Department of Lightweight Structures and Polymer Technology, Chemnitz University of Technology, Reichenhainer Straße 31/33, 09126 Chemnitz, Germany; ricardo.decker@mb.tu-chemnitz.de (R.D.), slk@mb.tu-chemnitz.de (L.K.)
*   Correspondence: rene.schmidt@informatik.tu-chemnitz.de; Tel.: +49-371-531-34671

Received: 29 June 2018; Accepted: 26 July 2018; Published: 31 July 2018

Featured Application: haptic input device for car interior or structural health monitoring crash relevant components.

Abstract: Achieving lightweight construction through only material substitution does not realize the full potential of producing a lightweight material, hence, it is no longer sufficient. Weight-saving goals are best achieved through additional function integration. In order to implement this premise for mass production, a manufacturing process for joining and forming hybrid laminates using a new tool concept is presented. All materials used are widely producible and processable. The manufactured cover of an automotive center console serves to demonstrate a human interface device with impact detection and action execution. This is only possible through a machine learning system, which is implemented on a small—and thus space- and energy-saving—embedded system. The measurement results confirm the objective and show that localization was sufficiently accurate.

Keywords: hybrid laminate; piezoceramic compound; sensor function; sheet metal forming; impact detection

1. Introduction

Laminates made of different materials, such as plastic and metal, are referred to as hybrid laminates. The combination of metal sheets with fiber reinforced plastics (FRPs) results in completely new property profiles, which are characterized by low weight combined with high specific rigidity and strength. This enables high degrees of lightweight construction. The insertion of the metal layers significantly improves the damage tolerance of the FRP, making the hybrid laminates predestined for use in the aviation industry. A well-known representative is GLARE (Glass Laminate Aluminum Reinforced Epoxy), a glass fiber reinforced epoxy resin aluminum foil laminate, which is the result of further development of the oldest known hybrid laminate, ARALL (aramid fiber reinforced epoxy resin aluminum foil laminate). The layer structure of GLARE consists of glass fiber prepreg layers combined with aluminum sheets. However, due to the use of the thermoset matrix material, which has a long curing time, and the complex plant technology required, the production costs for these hybrid components increase [1]. The challenge to reduce these costs was met in a joint project at Chemnitz University of Technology, and hybrid laminates with a thermoplastic matrix were developed under the brand names CAPAAL© (carbon fiber reinforced polyamide aluminum laminate) and CAPET© (carbon fiber reinforced polyetheretherketone titanium laminate) [2].

Another approach to decreasing the weight is the integration of additional functions in structural components. Piezoceramic elements, for example, can be injection-molded with electrically conductive polymers and used, for example, for structural monitoring in fiber reinforced plastics [3]. Similarly, piezoceramic fibers—which are inserted into microcavities produced by forming, and joined by forming technology—enable the monitoring of metal components [4]. However, the efficient and large-scale production of hybrid laminates with large-area integration of sensors for detection and localization of impacts and deformations in hybrid sheet metal structural components is not yet available.

The forming of hybrid laminates presents a particular challenge in order to avoid failure of the individual layers due to tearing, delamination, and wrinkling. Successful forming can therefore only be achieved by adequate temperature control [5]. Harhash [6] goes on to discuss the effect of the core thickness on the mechanical properties. Furthermore, the failure during deep drawing is described, as is the beginning of crack propagation. Further research on the forming of hybrid laminate can be found in [7]. The laminate is preheated and then formed in a heated tool. The exact description of the material properties at elevated temperature, especially near the melting temperature, is a great challenge for the finite element (FE) simulation. A possible approach is the determination of these characteristic values by a representative volume element [8]. Nondestructive testing of hybrid laminates is also intended to detect defects that have a direct effect on the forming properties [9].

One field of application for hybrid laminates is represented by impact detection. In this field of application, three main algorithm approaches have emerged. The first class includes algorithms detecting impacts by precomputed reference values [10,11]. These algorithms face the drawback of low adaptability, since each component has to be measured individually. The second class represents signal-theoretical approaches for impact detection relying on direction of arrival estimation. Determination of the direction of arrival is based on cross-correlation [12], which is used for impact detection and localization on aircraft wings [13]. The third class covers localization with machine learning methods. For this purpose, various signal properties are extracted and used as inputs for various machine learning training techniques [14]. The main difference between the last two described classes are the good results from the signal processing class, as long as the objects are well shaped; if the objects are strongly deformed and no mathematical model can be found, machine learning approaches are preferred. The number of machine learning algorithms rises constantly and covers a wide range of research and application fields. Therefore, [15] used the support vector machine, while [16] relied on the kernel extreme learning machine. The most commonly used techniques are neural networks [17,18], whereby the quality of the procedure always depends on the features used to represent the selected signal property. Haywood et al. [19] and LeClerc et al. [20], for example, used features generated by Fourier and Hilbert transform, accompanied by high computational costs. In contrast, [12] focuses on features with low computational effort. The machine learning methods and signal properties to use cannot be generically determined and depend on the respective application.

The rest of this work is organized as follows. First, a novel manufacturing process for hybrid laminates with sensor functionality is described, addressing a continuous process chain and guaranteeing constant quality. Subsequently, a digital signal processing chain is described, demonstrating that the manufacturing process is accurate, does not damage sensor functionality, and has practicable usability. Finally, a conclusion about the presented results is drawn.

2. Manufacturing of Hybrid Laminates with Sensor Functionality

For manufacturing of hybrid laminates with sensor functionality [21], a large-scale process chain was developed. This innovative process chain combines the polymer processing technologies of compounding and foil extrusion with the rolling and forming of metal parts. In the first subprocess (Figure 1, label 1), the starting materials are blended into a piezoceramic thermoplastic compound and extruded as a thin foil, which forms the active component of the hybrid laminate. Therefore, the polypropylene homopolymer Moplen HP501H (LyondellBasell, Rotterdam, the Netherlands) is functionalized with piezoceramic lead zirconate titanate (PZT) powder NCE 55 (Noliac A/S, Kvistgaard, Denmark) and carbon nanotubes (CNTs) NC7000 (Nanocyl S.A., Sambreville, Belgium). The optimal material composition and processing properties were determined in previous investigations. Hence, the piezoceramic foil consisted of 70 wt % PZT and 0.5 wt % CNTs [22,23]. A foil thickness of 250 μm provides the optimal conditions for joining (Figure 1, label 2) the piezoceramic foil with aluminum sheets. The surface of the aluminum metal sheet (EN AW-6082 T4 with a thickness of 0.5 mm) was treated before joining. Grinding and additional etching with sodium hydroxide has proven to be a reliable method (acceptable adhesion when close to a large-scale reaction). After joining the piezoceramic foil and the copper sheets, cutting was carried out by water jet. A special tool concept was developed for the forming process, in which the composite is heated, formed, and cooled in the tool (Figure 1, label 3).

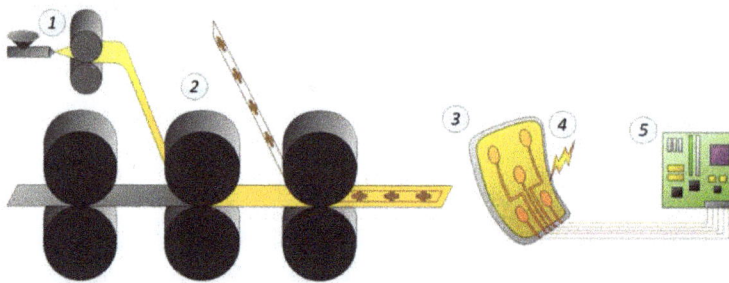

Figure 1. Process chain for mass production-enabled manufacturing of hybrid laminates: ① Foil extrusion, ② Joining, ③ Forming, ④ Polarization, ⑤ Signal Processing.

For continuous joining, the first tests were carried out on a pilot rolling mill [24]. After successful completion, a system was developed which allows a sample size with a width of up to 300 mm and a theoretically infinite length. The concept is shown in Figure 2. The system consists of a sheet metal inlet, which is also used for heating, and a sheet metal outlet, in which the finished composite cools down slowly. The actual joining process takes place directly after heating by infrared radiation (Figure 2, right A). After feeding the piezoceramic foil and (optionally) the copper strips, the composite is joined by the pressure of the rubber rollers. The thermoplastic melts in the contact area and forms a strong bond by adhesion. The process is monitored by thermocouples under the infrared radiators, and the temperature is measured by a pyrometer just before the rolls. The first tests showed that the parameters are not completely transferable from the pilot rolling mill. Due to the larger width of the sheets, the infrared emitters were arranged differently, thus causing better coupling and heating in the sheet. After a short parameter variation, it was possible to find the appropriate setting and produce hybrid laminates with sensor functionality.

The forming of the hybrid laminates was investigated in preliminary tests by V-bending and deep-drawing. The forming temperature, at which the brittle plastic does not tear or flow out, was particularly important. All the described preliminary work was carried out with a tool heated by heating cartridges and samples preheated in the furnace. The V-bending experiments showed that if the temperature is too low, the piezoceramic compound cracks. This is due to the fact that it is rather brittle due to the high proportion of ceramics. Only from a forming temperature of 100 °C did the piezoceramic compound not fail. With a further increase in temperature to 180 °C, the plastic flowed due to the reduction in viscosity. This was particularly evident at 180 °C, at which the piezoceramic compound was strongly thinned out and flowed laterally out of the sample (a more exact description is published in [24,25]). To further limit the forming temperature, DSC (differential scanning calorimetry) analysis of the piezoceramic compound was carried out. DSC confirmed the results of the preliminary tests: the melting starts at 110 °C and the maximum enthalpy of melting is reached at 167.8 °C, which also represents the melting point.

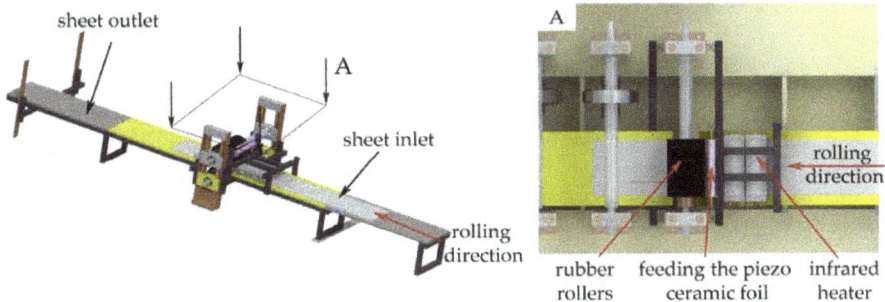

Figure 2. Continuous joining of aluminum sheets with piezoceramic foil and the top view (A).

The findings were transferred to the manufacture of a cover for the center console of an automobile. With this part, it should be possible to carry out different actions in the car by localizing impacts. A new tool concept was developed for the forming process. Figure 3 shows the overall structure of a servo-electric press. Instead of heating cartridges, a variothermal heater was used, which uses water to heats the tool to 190 °C and cools it down to 20 °C again. This occurs by two separate circuits, which are connected by valves. Furthermore, an oven for preheating was omitted, and the heating was placed directly in the press by means of infrared radiators (Figure 3, right B). The procedure is as follows: The sample is inserted into the sample holder and preheated by means of infrared radiation. The specimen holder can be moved and is pushed on a guide to the tool. The free programming of the servo-electric press allows initial bending of the outer cover geometry when the tool is closed in a path-controlled manner, and then switches to a force-controlled control system in order to deep-draw the inner contour with the drawing cushion. Upon completion of the forming process, the hot forming tool is rapidly cooled by the variotherm unit. Thus, the plastic solidifies, and the component can be removed afterward.

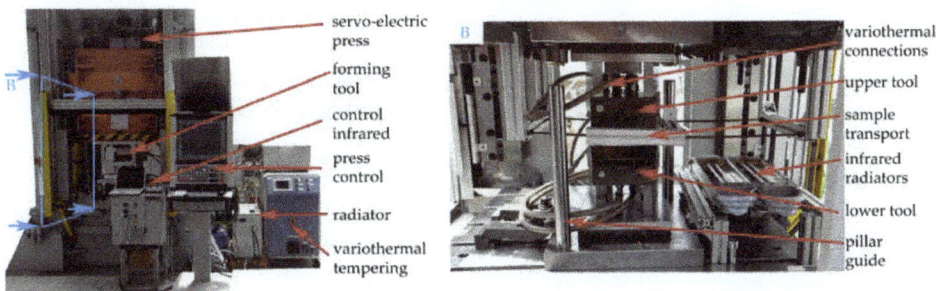

Figure 3. Variothermal forming tool in a servo-electric press and a detailed side view (B).

To adjust the process, a sample made only of aluminum was provided with thermocouples in order to record the working temperature (Figure 4). The rate of the temperature change was determined by the first derivative over time, which clearly showed the alternation of heating, holding, and cooling. The process steps described above are indicated in Figure 4.

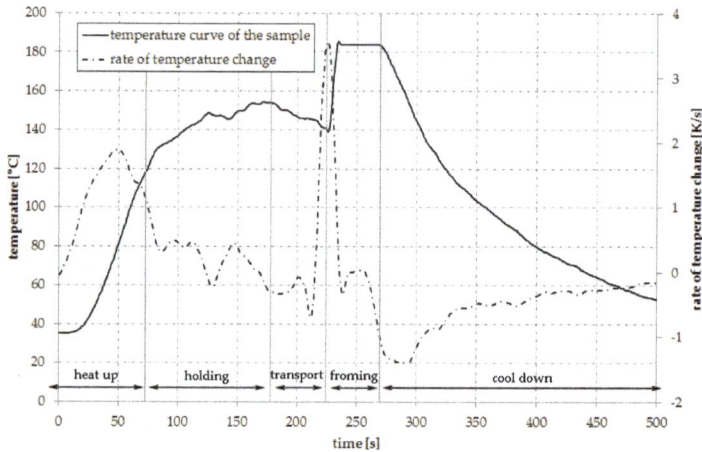

Figure 4. Temperature measurement of a sample during the process.

In the tests, the preheating temperature and the die temperature proved to be the most important parameters for defect-free forming. If these temperatures are set too low, tears and/or delamination occurs. Figure 5 shows the behavior if the temperature in the process is too high and the viscosity of the plastic is too low. It can be seen that the plastic flows out of the drawing gap, and thinning occurs in the middle. The parameters for the experiment were: preheating temperature infrared radiator 200 °C and tool temperature 170 °C. These parameters are identical to those from the test in Figure 4, and, when looking closely at the measured value, it can be determined that the temperature was above the melting temperature of the DSC analysis. After further tests, and by adjusting the parameters to the preheating temperature of the infrared radiator 180 °C and the tool temperature 150 °C, a component could be produced without thinning and melting, and without cracks. Two variants formed with these parameters are shown in Figure 6. The first variant is without copper strips (Figure 6a); it is mainly used for the individual application of electrodes. The second variant represents the complete mass production of a component with electrodes (Figure 6b). The results showed that the process is mastered and that, with the right parameters, a component can be produced without defects.

Figure 5. Formed hybrid laminates with sensor functionality at preheating temperature 200 °C and tool temperature 170 °C.

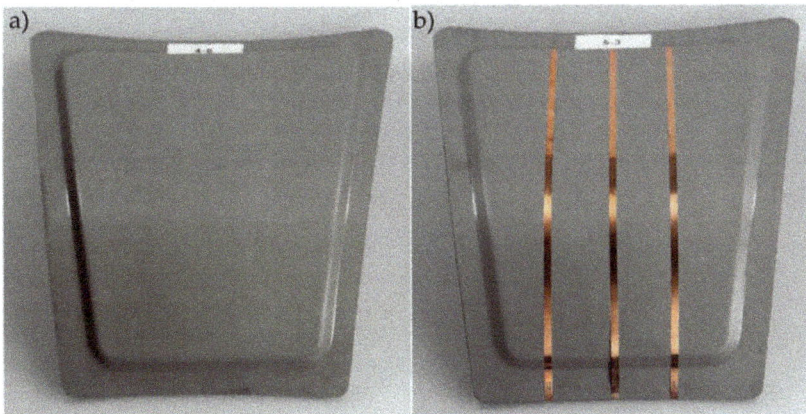

Figure 6. Formed hybrid laminates with sensor functionality in two versions: (**a**) without electrodes and (**b**) with electrodes at preheating temperature 180 °C and tool temperature 150 °C.

For activating the sensor effect of the hybrid laminate, a polarization process has to be conducted (Figure 1, label 4). Due to the extrusion, joining, and forming temperatures, which are higher than the Curie temperature of the PZT component, the piezoceramic particles depolarize during these processes. Therefore, the functionalized hybrid laminate must be polarized after the forming process. Based on previous experiments, the polarization process was conducted at elevated temperatures, between 100 °C and 120 °C [24]. In this temperature range, it is possible to polarize the piezoceramic layer during the cooling phase, after forming the hybrid part. By using this procedure, the residual heat of the forming process renders the reheating for polarization redundant. This approach has great potential for saving energy and resources, as well as for reducing cycle times during the fabrication process of multifunctional lightweight structures based on hybrid laminates.

In order to demonstrate real-world functionality, practical usability, and accuracy of the described manufacturing process, one center console, shown in Figure 6a, was equipped with four sensors to detect and localize the touch of a human finger. Therefore, the following section presents a novel measurement system architecture with integrated machine learning functionality.

3. Signal Processing

3.1. System Description

The sensor function of the hybrid laminate uses the piezoelectric effect. Mechanical loads on the multifunctional lightweight structure, e.g., impacts on the surface, induce displacements of the electrical dipoles within the piezoceramic particles. These displacements generate measurable electrical voltages between the copper electrodes and the aluminum sheet of the hybrid laminate, which can be further processed with the assistance of digital measuring methods. One possible application is represented by impact detection, which is used as an exemplary case for showing the functionality of the manufacturing process of the sensors, as well as the given practical usability. In our previous work, Ullmann et al. showed that, with the same sensors, a sufficiently high localization accuracy can be achieved on the basis of a Support Vector Machine (SVM) usage combined with time difference computation, although the test objects are mathematically difficult to describe [26]. However, for practical use, real-time functionality has to be ensured, just as energy efficiency must be confirmed. The previous solutions from [26] offer a solid basis for data collection and evaluation, but they are unsuitable for practical use. The main problem is represented by the used UART Interface, since the provided data rates are too small for the large amounts of data, leading to big buffers accompanied by high hardware requirements. The second main problem is that data processing on conventional PCs limits the practical use in cars or airplanes. For this reason, the process was transferred to an energy-efficient embedded system on chip (SoC) design, providing real-time capability, low power consumption, flexible application possibilities, and the accuracy necessary for real-world scenarios.

The basis for the design is a Zynq XC7Z010 SoC provided by Xillinx, representing a combination of FPGA and a Cortex A9 ARM processor connected by a standardized AXI interface, providing fast and highly flexible communication methods accompanied by high data throughput. The novel system architecture is depicted in Figure 7. As a test object, a center console with four sensors was used—three were placed in a vertical column and one was off-centered with a base distance of 5 cm. The object is stimulated by a light impact generated by contact from a human finger, leading to a visible voltage difference at the sensors. However, the sensor signals have a very low impedance, leading to insufficient measurement results by ADC usage. For data acquisition, the analog signals have to be stabilized energetically first. Afterwards, the signals must be transformed to the corresponding measuring range of the ADC. This task is performed by an electric circuit, which ensures the correct transformation of the analog signals to digital signals. The stabilized data are sampled and digitized using Xilinx's integrated XADC. The configuration and data sampling are handled by the ADC interface depicted in Figure 7. In the following steps, the data are smoothed by a moving average filter, and the necessary features are extracted and implemented on the FPGA. The resulting features are transferred to the SVM classification process realized in software executed on the ARM processor. Due to embedded Linux usage on the processor part, all arbitrary communication interfaces, like Ethernet, CAN, or UART, can be used to send the localization results to the most common possible applications. In the following, the necessary steps for impact detection are explained in detail.

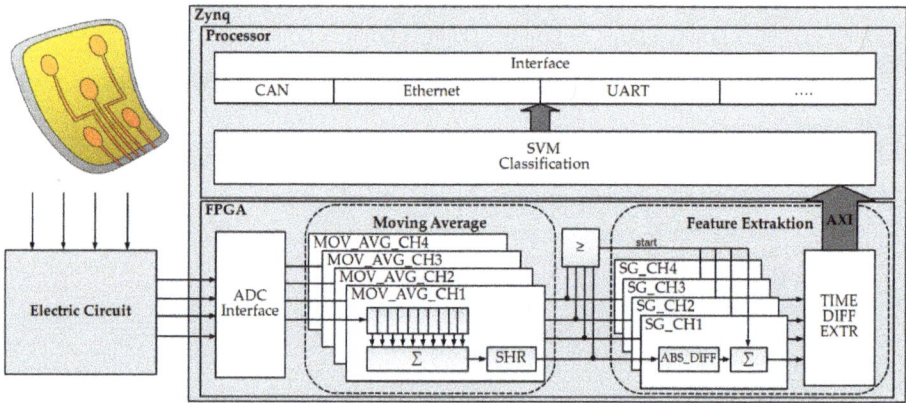

Figure 7. Novel system architecture for real-time processing on the basis of Zynq SoC (XC7Z010) with complete process flow.

3.2. Electric Circuit

The sensor signal generation is based on the piezo effect, leading to an increasing voltage on the sensor output. However, the sensor has no external power supply, resulting in voltage changing without generating new current. Consequentially, the sensor signal has too low an impedance to be measured with common ADCs. Therefore, the low impedance signal must be transformed into a high impedance signal to ensure the correct ADC conversion; additionally, the value range of the ADC must be secured. For this reason, the first process step is an impedance converter, which infuses additional energy to the signal without changing the voltage characteristic (cp. Figure 8). The amplifier used should not influence the sensor signal frequency, and it should work with the onboard voltage as a power supply. In the following steps, the measuring range of the XADC, from 0 V to 1 V, is ensured. Therefore, the voltage has to be shifted to the positive voltage range by applying an offset of 0.5 V, leading to an inversion of the signal determined by the characteristics of the used operational amplifier (LM324N). For this reason, the subsequent step inverts the signal again. In the last step, the limitation of the measurement range must be secured. This is realized by two limiters cutting the signal at less than 0 V and higher than 1 V. As a result, the signal is converted from a low impedance signal with an arbitrary range to a signal with high impedance shifted by 0.5 V, limited to the ADC measurement range.

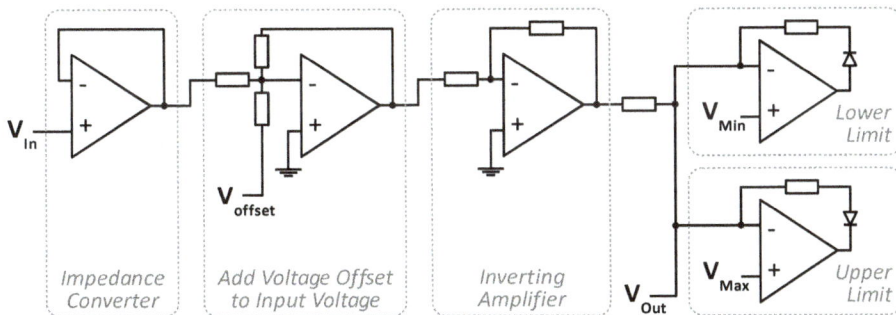

Figure 8. Electric circuit for transforming the low impedance input signal to high impedance signal, shifted by 0.5 V and limited to a 0–1 V ADC measurement range.

3.3. Signal Preprocessing

The preprocessing of the data starts with the data acquisition at the XADC, which converts the generated analog voltages of the electrical circuit into digital values, represented as the ADC interface in Figure 7. The XADC is a multichannel ADC with a theoretical maximum sampling rate of 1 MSps. Internally, the XADC consists of two synchronized ADCs with a sampling rate of 1 MSps and a configurable multiplexer used for switching the different measurement channels. For this reason, the number of used channels is not negligible, since it defines the final sample rate per measurement channel and can lead to wrong conversion results. In the described architecture, four channels are used, where two channels are mapped to ADC1 and the other two channels are mapped to ADC2, respectively, leading to a sample rate of 500 kSps on ADC1 and ADC2. To guarantee data consistency, ADC2 has been configured to work synchronously with ADC1, concluding in a final sample rate of 500 kSps, which defines the measurement window size and buffer sizes. Switching the channels and initial configuration of the XADC is guaranteed by the ADC interface.

A gradient-based algorithm is used as a feature extractor for the machine learning part (cp. Section 3.4.). The gradient calculation is sensitive to interferences, since a high-frequency noise signal is always accompanied by fast-changing amplitudes, leading to higher gradients. For this reason, the data are smoothed using a moving average filter. For this purpose, a shift register was implemented that satisfies the condition $n = 2^i$, with $i \in \mathbb{N}$ and shift register length n, thus efficiently implementing the division by a power of 2. The division by a power of 2 can be implemented by a shifting operation of i bits, represented as the SHR block in Figure 7. Since determining the sum for large measurement window sizes would require a lot of resources, the filter value m was calculated according to formula:

$$m(x)_t = \frac{1}{n}\left(\sum_{i=0}^{t-1}x(i) - \sum_{i=0}^{t-1-n}x(i)\right) \tag{1}$$

The first part of the equation can be implemented efficiently by continuously accumulating the incoming samples. The second part represents the same summation, but the signal is shifted by n samples in time. Therefore, the input signal is delayed for n samples by the shifting register and is subtracted continuously. This offers the advantage of linear scalability, while the arithmetic operation is reduced to a summation and subtraction, leading to complexity reduction and avoiding timing problems.

3.4. Feature Extraction and Impact Localization

The feature extraction is based on the determination of the summed gradient, which is displayed in Figure 7 as the SG block. The first step of the feature extraction is the parallel transfer on each separated channel. The signal filtered during preprocessing serves as continuous input signal. The first processing step is the gradient calculation, represented as the ABS_DIFF block in Figure 7. The gradient is defined as the absolute difference between two consecutive samples. The implementation is carried out by following equation:

$$g(m)_t = \sum_{i=1}^{t} |m(i) - m(i-1)| \tag{2}$$

In the process, the gradients from a defined start time $t = 0$ are added up successively. The required start time is generated by a threshold comparison. For this purpose, the amplitude of the filtered signals on all measurement channels is compared to a parameterizable threshold value, generating the start signal for summation when all amplitudes exceed the defined threshold.

The result is the sum of the absolute gradients at a specific point in time per the measurement channel depicted in Figure 9. The generated data are passed to the subsequent function block TIME_DIFF_EXTR. This block determines the times of the gradients when they exceed a specified threshold value. For this purpose, a counter is used as a time reference, which is incremented with each incoming sample. To prevent overflows and the associated false measured values, the counter is in the reset state until the summation of the gradients begins. This creates a direct start condition and implicitly defines the start of the time measurement. The counter value is stored each time the threshold value is exceeded for the first time. If all channels fall below the threshold value, the measurement is considered complete. The resulting time difference representing the signal shift on the different measurement channels is sent to the processing system via AXI interface, proceeding to the SVM classification process.

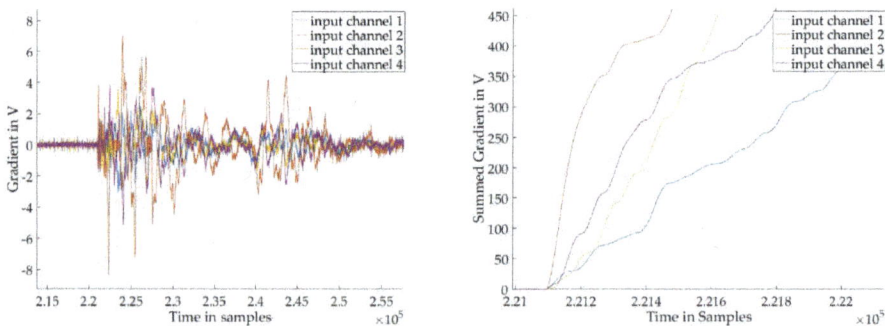

Figure 9. Example of calculated gradient (**left**) and summed gradient (**right**).

In the software, machine learning occurs by an SVM. For this reason, an SVM was realized on the basis of LIBSVM [27], which provides a high-performance cross-platform solution. To generate the SVM model, 20 measuring points on the center console were defined in a grid pattern. Twenty-six measurements were carried out at each of these measuring points, resulting in 520 measurements, of which 400 were used for training and 120 were assigned as the test data set. For evaluation purposes, the described design was implemented with a window size of 64 samples, an amplitude threshold 2 times higher than the maximum amplitude in idle mode, and a threshold for gradient comparison of 40 V, as identified by heuristic evaluation. To evaluate the SVM kernel functions, the linear, the polynomial, the radial base, and the sigmoid kernel were trained with the described training data set resulting in individual models, which were applied to the test data set. The results of the training and test phases are depicted in Figure 10. It is shown that the radial basis kernel provides very good training results, whereas the test results are not sufficient, while the linear kernel and the polynomial kernel provide less accurate training results but provide sufficient results on the test data set. On the other hand, the sigmoid kernel provides insufficient results in testing and training.

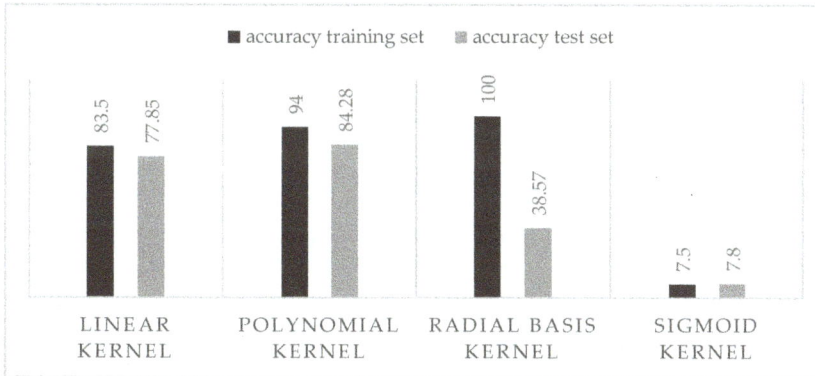

Figure 10. Comparison between different kernel functions for support vector machine (SVM) training with respect to accuracy on the training data set and test data set.

4. Conclusions

A new hybrid laminate with sensor functionality was presented. It offers the possibility of combining lightweight construction and functional integration into a component, and producing it in large quantities. For this purpose, it was necessary to develop a continuous joining technology which combines continuous piezoceramic foil and an aluminum sheet, and it guarantees a constant quality. This was achieved through previous research and further investigation.

A new heating concept and a new tool concept using variothermal tempering were developed for further processing. Both have been integrated into a servo-electric press to minimize cycle time and energy loss caused by transport. By measuring the temperature on a sample plate, the process was verified. Good parts of high quality were produced with appropriate parameters.

In further investigations, the parts will be scanned by an optical measuring system and compared with the target geometry. The acquired information about thinning and springback will form the basis of adjusting the FE simulation and further improving the process. The polarization behavior of hybrid laminates with modified PZT components, as well as the development of tool concepts for in-line polarization processes, are still part of ongoing research activities.

In the signal processing part, an electrical circuit was implemented to amplify the low impedance piezoelectric effect signal and to ensure the correct measuring range. On the other hand, a full pipelined FPGA structure was presented, guaranteeing real-time processing. On the basis of the novel SoC architecture, the feature determined in the FPGA can be passed directly to the ARM processor, which classifies the features using an SVM, which had 94% accuracy on the training set and 84% on the testing set. With a power consumption of approximately 1.7 W, the presented solution is also very efficient in terms of energy consumption, enabling usage in mobile devices. However, although the used data set represents a valid basis for ensuring the evidence of feasibility, additional human-specific parameters, such as touch pressure, touch length, and characteristics of human fingers, have to be evaluated in further research before commercial use.

In summary, a novel manufacturing process for the production of piezo-based hybrid laminates was described. Further processing of the laminate—forming and in-line polarization—enable multifunctional structures, e. g., automotive center consoles with integrated touch functionality. The input gestures are processed by machine learning methods on an embedded system. The accuracy of the procedure was presented by an energy-efficient and real-time SoC solution, which precisely localized points of touch by human fingers. This demonstrates the practicability and usability of the hybrid laminates and the signal processing.

Author Contributions: Writing—aoriginal draft, R.S., A.G. and R.D.; Writing—review & editing, V.K., W.H., D.L. and L.K.

Funding: Deutsche Forschungsgemeinschaft: EXC 1075; Deutsche Forschungsgemeinschaft: GRK 1780/1; Deutsche Forschungsgemeinschaft: SFB/TR 39.

Acknowledgments: We gratefully acknowledge the cooperation of our project partners and the financial support of the DFG (Deutsche Forschungsgemeinschaft) within the Collaborative Research Center/Transregio 39 "PT-PIESA", the Federal Cluster of Excellence EXC 1075 "MERGE" and GRK 1780/1 CrossWorlds.

Conflicts of Interest: The authors declare no conflict of interest.

References

1. Vlot, A.; Gunnink, J.W. *Fibre Metal Laminates an Introduction*; Kluwer Academic Publishers: Dordrecht, The Netherland, 2001; ISBN 1-4020-0038-3.
2. Wielage, B.; Nestler, D.; Steger, H.; Kroll, L. CAPAAL and CAPET—New Materials of High-Strength, High-Stiff Hybrid Laminates. In *Integrated Systems, Design and Technology 2010*; Springer: Berlin, Germany, 2011; pp. 23–35.
3. Kroll, L.; Walther, M.; Nendel, W.; Heinrich, M.; Tröltzsch, J. Initial Stress Behaviour of Micro Injection-Moulded Devices with Integrated Piezo-Fibre Composites. In *Integrated Systems, Design and Technology 2010*; Springer: Berlin, Germany, 2011; pp. 109–120.
4. Müller, M.; Müller, B.; Hensel, S.; Nestler, M.; Jahn, S.F.; Wittstock, V.; Schubert, A.; Drossel, W.G. Structural integration of PZT fibers in deep drawn sheet metal for material-integrated sensing and actuation. *Procedia Technol.* **2014**, *15*, 659–668. [CrossRef]
5. Kräusel, V.; Graf, A.; Nestler, D.; Jung, H.; Arnold, S.; Wielage, B. Forming of new thermoplastic based fibre metal laminates. In Proceedings of the 3rd Global Conference on Materials Science and Engineering (CMSE 2014), Shanghai, China, 20–23 October 2014; pp. 40–46.
6. Harhash, M.; Carradó, A.; Palkowski, H. Mechanical properties and forming behaviour of laminated steel/polymer sandwich systems with local inlays—Part 2. *Compos. Struct.* **2017**, *160*, 1084–1094. [CrossRef]
7. Mosse, L.; Compston, P.; Cantwell, W.J.; Cardew-Hall, M.; Kalyanasundaram, S. Stamp forming of polypropylene based fibre–metal laminates: The effect of process variables on formability. *J. Mater. Process. Technol.* **2006**, *172*, 163–168. [CrossRef]
8. Graf, A.; Lachmann, L. Modellierung der Umformung von hybriden Schichtverbunden mit thermoplastischer Matrix. In Proceedings of the Sächsische Fachtagung Umformtechnik (SFU), Dresden, Germany, 27–28 November 2013; TU Dresden: Dresden, Germany, 2013; pp. 43–52.
9. Neugebauer, R.; Kräusel, V.; Graf, A. Process Chains for Fibre Metal Laminates. *Adv. Mater. Res.* **2014**, *1018*, 285–292. [CrossRef]
10. Tracy, M.; Chang, F.K. Identifying impacts in composite plates with piezoelectric strain sensors, part I: Theory. *J. Intell. Mater. Syst. Struct.* **1998**, *9*, 920–928. [CrossRef]
11. Wölfinger, C.; Arendts, F.J.; Friedrich, K.; Drechsler, K. Health-monitoring-system based on piezoelectric transducers. *Aerosp. Sci. Technol.* **1998**, *2*, 391–400. [CrossRef]
12. Ribay, G.; Catheline, S.; Clorennec, D.; Ing, R.K.; Quieffin, N.; Fink, M. Acoustic impact localization in plates: properties and stability to temperature variation. *IEEE Trans. Ultrason. Ferroelectr. Freq. Control* **2007**, *54*, 378–385. [CrossRef] [PubMed]
13. Zhao, X.; Gao, H.; Zhang, G.; Ayhan, B.; Yan, F.; Kwan, C.; Rose, J.L. Active health monitoring of an aircraft wing with embedded piezoelectric sensor/actuator network: I. Defect detection, localization and growth monitoring. *Smart Mater. Struct.* **2007**, *16*, 1208. [CrossRef]

14. Park, B.; Sohn, H.; Olson, S.E.; DeSimio, M.P.; Brown, K.S.; Derriso, M.M. Impact localization in complex structures using laser-based time reversal. *Struct. Heal. Monit.* **2012**, *11*, 577–588. [CrossRef]
15. Fu, H.; Xu, Q. Locating impact on structural plate using principal component analysis and support vector machines. *Math. Probl. Eng.* **2013**, *2013*. [CrossRef]
16. Fu, H.; Vong, C.M.; Wong, P.K.; Yang, Z. Fast detection of impact location using kernel extreme learning machine. *Neural Comput. Appl.* **2016**, *27*, 121–130. [CrossRef]
17. Tsou, P.; Shen, M.H. Structural damage detection and identification using neural networks. *AIAA J.* **1994**, *32*, 176–183. [CrossRef]
18. Jones, R.T.; Sirkis, J.S.; Friebele, E.J. Detection of impact location and magnitude for isotropic plates using neural networks. *J. Intell. Mater. Syst. Struct.* **1997**, *8*, 90–99. [CrossRef]
19. Haywood, J.; Coverley, P.T.; Staszewski, W.J.; Worden, K. An automatic impact monitor for a composite panel employing smart sensor technology. *Smart Mater. Struct.* **2004**, *14*, 265. [CrossRef]
20. LeClerc, J.R.; Worden, K.; Staszewski, W.J.; Haywood, J. Impact detection in an aircraft composite panel—A neural-network approach. *J. Sound Vib.* **2007**, *299*, 672–682. [CrossRef]
21. Kräusel, V.; Graf, A.; Heinrich, M.; Decker, R.; Caspar, M.; Kroll, L.; Hardt, W.; Göschel, A. Development of hybrid assembled composites with sensory function. *CIRP Ann.* **2015**, *64*, 25–28. [CrossRef]
22. Graf, A.; Decker, R.; Kräusel, V.; Landgrebe, D.; Kroll, L. Suitable process for mass production of hybrid laminates with sensor functionality. In Proceedings of the International Conference on Hybrid Materials and Structures, Bremen, Germany, 18–19 April 2018; pp. 140–145.
23. Decker, R.; Heinrich, M.; Tröltzsch, J.; Rhein, S.; Gebhardt, S.; Michaelis, A.; Kroll, L. Development and Characterization of Piezo-active Polypropylene Compounds Filled With PZT and CNT. In Proceedings of the 5th Scientific Symposium CRC/Transregio 39, Dresden, Germany, 14–16 September 2015; pp. 59–64.
24. Ullmann, F.; Decker, R.; Graf, A.; Kräusel, V.; Heinrich, M.; Hardt, W.; Kroll, L.; Landgrebe, D. Continuous manufacturing of piezoceramic hybrid laminates for functionalised formed structural components professorship for forming and joining. *Technol. Light. Struct.* **2017**, *1*, 1–13.
25. Graf, A.; Kräusel, V.; Landgrebe, D.; Decker, R.; Kroll, L. Joining and forming of hybrid assembled composites with sensory function. In Proceedings of the Euro Hybrid Materials and Structures, Kaiserslautern, Germany, 20–21 April 2016; pp. 118–124.
26. Ullmann, F.; Hardt, W.; Zhmud, V. Machine learning algorithms for impact localization on formed piezo metal composites. In Proceedings of the 2017 International Siberian Conference on Control and Communications (SIBCON), Astana, Kazakhstan, 29–30 June 2017; IEEE: Piscataway, NJ, USA, 2018; pp. 1–5.
27. Chang, C.C.; Lin, C.J. LIBSVM: A library for support vector machines. *ACM Trans. Intell. Syst. Technol.* **2011**, *2*, 27. [CrossRef]

MDPI

St. Alban-Anlage 66

4052 Basel

Switzerland

Tel. +41 61 683 77 34

Fax +41 61 302 89 18

www.mdpi.com

Applied Sciences Editorial Office

E-mail: applsci@mdpi.com

www.mdpi.com/journal/applsci